6th International Conference on Ship and Offshore Technology (ICSOT Indonesia 2019)

Developments Marine Design, Construction and Operation

Semarang, Indonesia
25-26 November 2019

Editors:

Good Rindo
Muhammad Iqbal
Berlian Arswendo Adietya

ISBN: 978-1-7138-3036-8

Printed from e-media with permission by:

Curran Associates, Inc.
57 Morehouse Lane
Red Hook, NY 12571

Some format issues inherent in the e-media version may also appear in this print version.

Printed with permission by Curran Associates, Inc. (2021)

For permission requests, please contact The Royal Institution of Naval Architects
at the address below.

The Royal Institution of Naval Architects
8-9 Northumberland Street
London, WC2N 5DA
United Kingdom

Phone: 020 7235 4622
Fax: 020 7259 5912

publications@rina.org.uk

Additional copies of this publication are available from:

Curran Associates, Inc.
57 Morehouse Lane
Red Hook, NY 12571 USA
Phone: 845-758-0400
Fax: 845-758-2633
Email: curran@proceedings.com
Web: www.proceedings.com

p-ISSN :2598-9006

Proceedings of International Conference on Ship and Offshore Technology

"Developments Marine Design, Construction and Operation"

Vol. 2, November 2019

Organized by :

Diponegoro University

Royal Institution of Naval Architects (RINA)

November, 25th - 26th 2019
Diponegoro University (UNDIP) - Semarang
Indonesia

Proceedings of International Conference on Ship and Offshore Technology

"Developments Marine Design, Construction and Operation"

November, 25th – 26th 2019
Diponegoro University (UNDIP) - Semarang
Indonesia

Vol. 2, November 2019

Published by :
Department of Naval Architecture

Proceedings of International Conference on Ship and Offshore Technology

"Developments Marine Design, Construction and Operation"

Publisher : Department of Naval Architecture
Publication frequency : 2 years
p-ISSN : 2598-9006

Welcome Message From Chair

On behalf of the Committee, it is our great pleasure to welcome you to the sixth International Conference on Ship and Offshore Technology – Indonesia (ICSOT Indonesia 2019). This year, the conference is jointly organized by Diponegoro University (UNDIP) and the Royal Institute of Naval Architects (RINA).

The conferences theme is "Developments Marine Design, Construction, and Operation". Although the theme is limited to developments marine design, construction, and operation, but the theme can be related and developed to other aspects such as hydrodynamics, marine safety, environment, material, construction aspects, maritime logistics, fishing vessel, etc. The program committee accepted 14 of those papers to be presented in this conference.

We are honored to have outstanding speakers. And also, special thanks to the Keynote Speakers: Prof. Dr. Hirotada Hashimoto, and Prof. Dr. Ir. Muljo Widodo Kartidjo. We appreciate to the Royal Institute of Naval Architects (RINA) as supporting organization in this conference, especially Vice President RINA, Prof. Ir. I Ketut Aria Pria Utama, MSc, PhD, Ceng, FRINA.

We welcome you to Semarang, and we very much hope you enjoy the conference.

Dr. Eng. Deddy Chrismianto, ST, MT
Chair of ICSOT Indonesia 2019

Table Of Contents

Organizing Commite

International Steering Committee (ISC)

Chair : Dr. Eng. Deddy Chrismianto, ST, MT (Diponegoro University, Indonesia)

Co-chair : Mr. Trevor Blakeley (RINA Chief Executive)

Members :

Prof. Richard Birmingham (Newcastle University)

Prof. Atilla Incecik (University of Strathclyde)

Prof. Jeom-Kee Paik (Pusan National University)

Prof. R A Shenoi (University of Southampton)

Prof. Nicholas Hutchins (University of Melbourne)

Prof. Takeshi Shinoda (Kyushu University, Japan)

Prof. Ru-Min Chao (National Cheng-Kung University, Taiwan)

Prof. Adi Maimun bin Abdul Malik (Universiti Teknologi Malaysia)

Prof. Yanuar (UI)

Prof I Ketut Aria Pria Utama (ITS)

Dr. Eng. Ahmad Fauzan Zakki, ST, MT (Diponegoro University, Indonesia)

Dr. Daeng Paroka (University of Hasanuddin)

Dr. Marcus Tukan (University of Pattimura)

Local Organizing Committee (LOC)

Reviewer : Dr. Eng. Deddy Chrismianto, ST., MT

Dr. Eng. Ahmad Fauzan Zakki, ST., MT

Dr. Eng. Hartono Yudo, ST., MT

Dr. Eng. Andi Trimulyono, ST., MT

Dr. Eng. Samuel, ST., MT

Muhammad Iqbal, ST, MT

Editor : Good Rindo, ST, MT

Muhammad Iqbal, ST, MT

Berlian Arswendo, ST, MT

ICSOT 2019 Agenda

Monday, 25th November 2019

Time	Agenda
07.30 - 08.30	Registration
08.30 - 09.00	Welcome Remark
09.00 - 09.15	Opening Speech (Vice President Asia Region : Prof. Ir. Ketut Aria Pria Utama, M.Sc, PhD, C. Eng, FRINA
09.15 - 09.30	Greeting of Dean of Engineering Faculty
09.30 - 09.40	Opening Ceremony ICSOT 2019
09.40 - 10.00	Coffee Break
10.00 - 10.30	Keynote Speaker 1 : Prof. Dr. Eng Hirotada Hashimoto (Kobe University, Japan)
10.30 - 12.00	**1st Session of Conference (3 presenters)**
12.00 - 13.00	Lunch
13.00 - 14.30	**2nd Session of Conference (3 presenters)**
14.30 - 15.00	Coffee Break
15.00 - 16.00	**3rd Session of Conference (2 presenters)**
18.30 - 20.00	Dinner

Tuesday, 26th November 2019

Time	Agenda
08.00 - 09.00	Registration
09.00 - 09.30	Keynote Speaker 2: Prof. Mulyo Widodo Kartidjo (Mechanical Engineering, ITB)
09.30 - 10.00	Coffee Break
10.00 - 11.30	4th Session of Conference (3 presenters)
11.30 - 13.00	Lunch
13.00 - 14.30	5th Session of Conference (3 presenters)
14.30 - 15.00	Closing Ceremony

Technical Program Schedule

No.	Title	Time	Session
1.	A Review State Of The Art Crew Boat Improving Fuel Efficiency Based On Operational Data Motion Using Artificial Intelligence	10.30 - 11.00	1st Session of Conference
2.	Experimental And Numerical Study Of Ship Resistance Due To Variation Of Hull Vane Positioning In The Longitudinal Direction	11.00 - 11.30	1st Session of Conference
3.	Numerical Investigation Into The Pressure Distribution And Form Factor Effect Of Slender body Catamaran	11.30 - 12.00	1st Session of Conference
4.	Study On The Environmental Condition Of Indonesia's Ports Anchoring Area	13.00 - 13.30	2nd Session of Conference
5.	The Optimization Of Anchor Equipment Due To The Specific Anchoring Area	13.30 - 14.00	2nd Session of Conference
6.	The Hull Girder Strength Analysis Due To Equipment Load Under Longitudinal Bending	14.00 - 14.30	2nd Session of Conference
7.	Geometry Optimization Of Centre Bulb To Reduce Wave Resistance On Catamaran Ship	15.00 - 15.30	3rd Session of Conference
8.	Toward Green And Sustainable Ship Recycling Industry In Indonesia	15.30 - 16.00	3rd Session of Conference
9.	A Model Scale Of Unsinkable Small Passenger Speed-Boat	10.00 - 10.30	4th Session of Conference
10.	The Development Of Intact Stability Criteria : The Work On Small Ship Up To 24 M In Operate In Indonesian Waterways	10.30 - 11.00	4th Session of Conference
11.	Application Of The Second Generation Intact Stability Criteria To An Indonesian Ro-Ro Ferry Supported By Model Experiment	11.00 - 11.30	4th Session of Conference
12.	Stability Assessment Of Hatch Coverless River-Sea Cargo Ship	13.00 - 13.30	5th Session of Conference
13.	Study On The Alternative Method Of Non-Destructive Testing For Fibre Reinforced Polymer	13.30 – 14.00	5th Session of Conference
14.	Comparative Study Of Straightening Method Using Oxy-Acetylene Carburizing With Tungsten Inert Gas On Aluminium 5083-H116	14.00 - 14.30	5th Session of Conference

A REVIEW OF THE STATE OF THE ART OF IMPROVING FUEL EFFICIENCY USING ARTIFICIAL INTELLIGENCE BASED ON FULL-SCALE OPERATIONAL DATA

S Riyadi, I K A P Utama, W D Aryawan, Institut Teknologi Sepuluh Nopember (ITS), Indonesia
R Rulaningtyas, Universitas Airlangga (UNAIR), Indonesia
G Thomas, University College London (UCL), United Kingdom

SUMMARY

Reducing crew boat fuel consumption is of the highest concern. Not only since operational data highlights that the cost of fuel is about 60% of total operational costs, but also due to the fossil fuel contribution to green house gas emissions which tend, to increase is not taken into consideration. Therefore, the use of a system to understand and provide the opportunity to control energy usage is highly recommended. This paper describes the approach of reducing fuel consumption using accurate estimates of wind, wave, and sea current effects, together with data collection from actual ship operations. Unfortunately the implication of such equipment to fill all data is expensive and complicated. Therefore, data analysis is needed together with the combination of engine records in order to fine art the improvment of fuel efficiency. In order to solve the problem, such solutions may to minimize data collection and will be equipt with motion sensor in existing SHIMOS® system. The current paper describes a step by step process of solving the above problem, together with the use of artificial intellience (AI) methods with adequate data collated from published literature.

NOMENCLATURE

[Symbol]	[Definition] [(Unit)]
AI	Artificial Intellience
ANN	Artificial Neural Network
DSS	Decision Support System
GPS	Global Positioning System
RPM	Rotational per Minute of Engine

1. INTRODUCTION

Reducing crew boat fuel consumption is of the highest concern. Not only since operational data highlights that the cost of fuel is about 60% of total operational costs, but also due to the fossil fuel contribution to green house gas emissions. Therefore, the use of a system to understand and provide the opportunity to control energy usage is highly recommended.

In actual practical that case above can solve with 2 kind of method. Achived by design and operational scheme [1]. Design achivment focus in optimum engine, vessel resistance, vessel weight, and vessel propeller[9]. And operational achivment focus to vessel speed, routing, hull and propeller, and engine maintenance. In detail in operational vessel other explain above, another is rotational per minute of engine RPM, mean draft, trim, cargo quantity, wind effect, and sea effect [2][3].

In ship operations the use of big data on the results of recording sensors mounted on ships includes:

- Routes, with tracking and GPS technology, actual ship routes are obtained so that the actual distance traveled within the route is obtained.
- Fuel consumption, the engine is currently equipped with a monitor of actual fuel requirements, if the engine does not yet have the technology, flowmeter can also be installed to monitor and record fuel requirements.
- Loading conditions, the load is very influential from the existing draft ship including trim conditions. So that the displacement accuracy is obtained as a reference in the estimation of resistance which will affect the fuel demand.
- Enviroment, environmental factors greatly influence the efficiency of fuel use. The ship's trajectory there are factors that influence the wind speed and direction, although it does not directly affect. The biggest factor is the existing waves. If there is bad weather so the waves are very high it will use up fuel so that the ship can run according to the target.

The data above so that it can be analyzed using statistics for the efficiency of propulsion [4][5][10]. Development of artificial intelligence on operational data is also applied so that energy efficiency is obtained [3]. Optimization of ship trim also influences the amount of resistance which ultimately efficiency [6]. The effect of the pattern of ship operating factors also greatly influences the performance of fuel requirements.

The opproach to improving fuel consumption in this study base on full scale operational data with case crew boat vessel. A data operational will carry out from SHIMOS®, installed in vessel [7]. The equipment for monitoring is limited. For future will be equip with motion measurement, the function is to record enviroment force in implication to the vessel.

Processing data from SHIMOS® with motion equipment as input, neuron in hidden layer, and output is improving fuel consumption. Artificial intelegence will hand for it.

2. CREW BOAT CASE STUDY

The ship in the case study is a crew boat that has the principal activity to regularly deliver special personnel to the rig platform. The boat has two main engines with waterjet propulsion system and can achieved a speed of 35 knot, so it is defined as a the high speed vessel.

Pothograph of crew boat it become case study shown in Figure 1. And for detail shown in table 2 explains the main characteristics of the crew boat that will be used in the study case.

Figure 1. The crew boat Display tracking rute and Position from SHIMOS

Table 2. Pricipal Dimension of Analysed Vessel

Princapal Dimension		Ship	
Length *Over All*	(LOA)	19.50	m
Breadth	(B)	4.50	m
Draft	(T)	0.95	m
Speed max	(Knot)	35.0	Kn
Engine Power	(Hp)	2 x 880	Hp
Deadweight	(Ton)	12.56	ton

3. SHIP MONITORING VESSEL (SHIMOS)®

To get data on the results of vessel operations being reviewed, there is On-board data monitoring device inside the ship. SHIMOS® is product of development from Orela shipyard to monitoring vessel, and used by PT. Pelayaran Nasional Ekalya Purnamasari. The data for the study are global positioning system (GPS) and fuel consumption from the main engine [8]. The equipment used to determine fuel consumption is a flowmeter. Which will record the amount of fuel flowing towards the ship's engine. The flowmeter data will be sent to the integrated computer system inside the ship, including the vessel position data obtained from GPS as shown in Figure 3.

Based on the GPS receiver. Data parameters used in SHIMOS® are the position of the ship, the speed of the ship over the ground via GPS, the direction of the ship's course, also the engine rpm data is obtained from the ship engine panel on the engine speed sensor.

Figure 3. Display tracking route and position from SHIMOS®

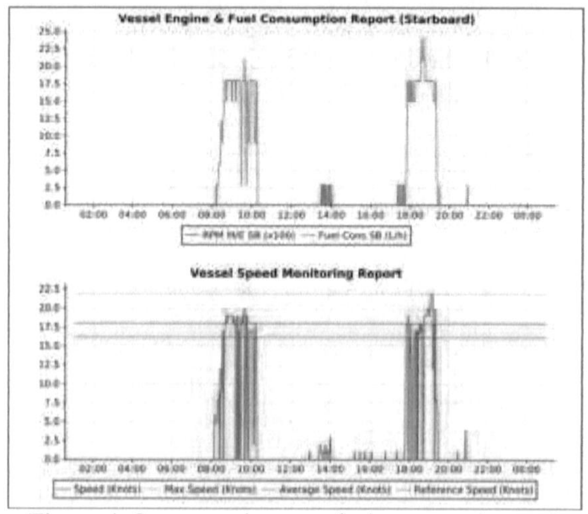

Figure 4. Output engine rpm, fuel consumption, and speed from SHIMOS®

The raw data, in the form of text obtained from the results of records sent from SHIMOS®, are processed into representative data. One of the dashboard outputs on the monitoring website can be displayed on SHIMOS® monitored vessel data. On the screen can be seen the condition of the speed of the ship and the crew boat trip. So that the data processing can be used as data material for evaluation related to the performance of the operating vessel. In Figure 4 shows a record display of ship speed, engine rpm and fuel consumption. Including processing average ship speed.

In future records, the ship will be equipped with a motion sensor. So the response from the surrounding environment (wind, ocean currents, waves, etc.) to the speed will be recorded as motion. the motion sensor that will be used will be discussed further in another section.

4. SHIP MOTION MEASUREMENT

Motion vessel measurements on ships using low cost sensors can be used as a measurement of the ship's response to the sea state and other environments [9].

There are several low cost sensors that can be used to measure the motion with a variety of support platforms such as Arduino, Rapberry, and Gadgeteer. This work used Rasbeery PI and sensor module GY-85 with 9 axis degree of freedom IMU sensor. This consist triple-axis gyroscope, triple-axis accelerometer, and triple-axis digital compass. Sample output is shown in figure 5.

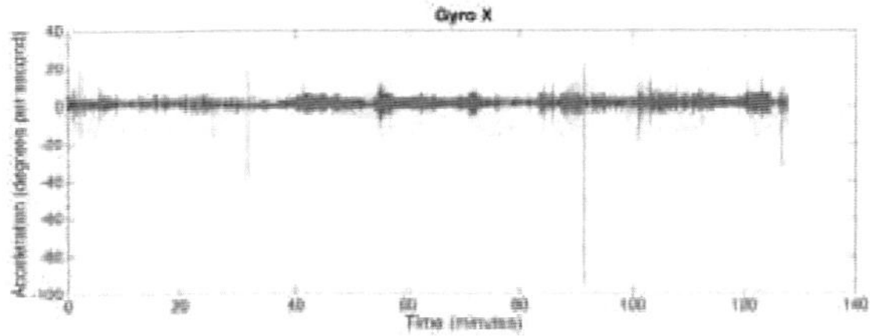

Figure 5. Sample output of Gyro X.

The sensor that will be used as a motion vessel measurement tool with the Android platform is the BNO055 Intelligent 9-axis absolute orientation sensor type model. they have integrating a triaxial 14-bit accelerometer, a triaxial 16-bit gyroscope with a range of ±2000 degrees per second, a triaxial geomagnetic sensor and a 32-bit cortex M0+ microcontroller. The accelerometer features ranges ± 2 g / ± 4 g / ± 8 g / ± 16 g, Gyroscope features ranges switchable from ± 125°/s to ± 2000°/s, and Magnetometer features magnetic field range typical ± 1300 μT(x-,y-axis); ± 2500 μT(z-axis) [10].

The motion sensor BNO055 will be connected and fit in SHIMOS system, including output data which will be further processed.

5. ARTIFICIAL INTELLIGENCE TO IMPROVING FUEL EFFICIENCY

In a paper written by E. Bal Besikci [3], they used to anArtificial Neural Network ANN and Combination Decision Support System DSS to investigate the influence of ship operation on energy efficiency. The parameters for decision making was conducted using seven important factors: (1) Ship speed, (2) Engine revolution per minute (RPM), (3) Mean draft, (4) Trim, (5) Cargo quantity on board, (6) Wind, and (7) Sea effect. These are used as input in the network training.

The proposal of ANN model shown in Figure 10. This method based on back propagation learning algorithm applied with one hidden layer. The NN architecture is 7-12-1 for fuel prediction represented number of input, neuron in hidden layers, and output.

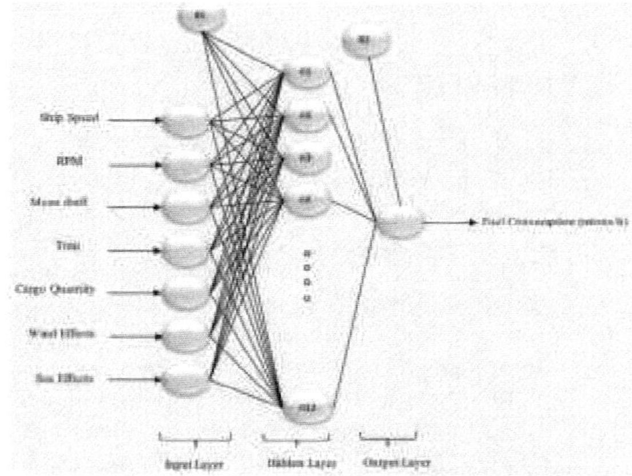

Figure 6. Schematic of ANN structure [3]

From statistic of fuel consumption already get form data was comsumed by the ship. The training and validation of ANN model for obseeved adn validation. Fuel consumption based on trained ANN parameter used weight and biases as function ship seven (7) parameter. Confident of result ANN need validation and benchmarking, used compare with multiple regression analysis MR.

Improving fuel consumption uses decision support system DSS needed for ship operator and advantage developed ANN. This method provides a strategic approach to make operational ship consider regarding both the economical and enviroment aspect.

Further work for aplication of this paper reference will be adopt. So all operational data will be filter to improving fuel consumption [11]. Paramater important will be change uses (1) ship speed, (2) Engine revolution per minute (RPM), (3) Average condtion from gyroscope, (4) Motion record from gyroscope. DSS aplied too with consider result fuel consumption for economical dan enviroment.

6. CONCLUSION

Improving fuel consumption to reduce it, will be carried out to decision for crew boat operational. The existing operation with monitoring use SHIMOS® and future will equipt with motion sensor and process with artificial intelegence.

Process of decision of artificial intelegence needed training with data operational and motion of vessel. Finally future output will be consider for operate the crew boat rpm engine to find improving reduction fuel consumption. Output method is build mode of operation to fits with operational actual.

7. ACKNOWLEDGEMENT

Authors wished to thanks PT. Pelayaran Nasional Ekalya Purnamasari to providing data vessel monitoring.

8. REFERENCES

1. K. K. KEE, B. Y. LAU SIMON, K. H Y. RENCO, 'Prediction of ship fuel consumption and speed curve by using statictical method', The 7th International Conference on Computer Science and Computational Mathematics (ICCSCM), 2018.

2. W. GORSKIL, T. A. GERIGK, and Z. BURCIU, 'The influence of ship operational parameters on fuel consumption', Scientific Journals Zeszyty Naukowe, 36(108) z., ISSN 1733-8670, 2013.

3. BAL BESIKCI, E., ARSLAN, O. TURAN, O. OLCER, 'An artificial neural network based decision support system for energy efficient ship operations', World Maritime University, PO Box 500, SE-20124 Malmö, Sweden, 2016.

4. J. P. PETERSON, D. J. JACOBSEN, O. WINTHER, 'Statistical modelling for ship propulsion efficiency', 2011.

5. D.G TRODDEN, A.J MURPHY, K. PAZOUKI, J SARGEANT, 'Fuel usage data analysis for efficient shipping operations', Elsevier Ocean Engineering 110 (75-84), 2015.

6. K. ZEYLAN, S. NAS, 'A study on the effects on trim optimatisation on ship resistace of a sub-panamax type container vessel', The Second Global Conference on Innovation in Marine Technology and the Future of Maritime Transportation, Bodrum, Muğla, TURKEY, 24-25 October 2016.

7. ORELA SHIPYARD IT TEAM, 'Guide and operational SHIMOS', Guide book, Surabaya Indonesia, 2016.

8. N. O. ABANKWA, S. J. JOHNSTON, M. SCOTT, & S. J. COX, 'Ship motion measurement using an inertial measurement unit', Research Gaye, DOI: 10.1109/WF-IoT.2015.738908, 2015.

9. A. F. MOLLAND, S. R. TURNOCK, D. A. HUDSON, and I. K. A. P UTAMA, 'Reducing Ship Emissions: A Review of Potential Practical Improvement in The Propulisve Effieciency of Future Ships', International Journal of Maritime Engineering, April 2014.

10. BOSCH SENSORTEC, 'BNO055 Intelligent 9-axis absolute orientation sensor', Data sheet, Document number BST-BNO055-DS000-12, Robert Bosch GmbH, Germany, 2014.

11. T. ANAN, H. HIGUCHI, and N. HAMDA, 'New Artificial Intellegence Technology Improving Fuel Efficient and Reducing CO2 Emmision of Ships through use of Operational Big Data', Fujitsu Sci. Tech. J., Vol. 53, No. 6, pp.23-28, October 2017.

9. AUTHORS BIOGRAPHY

S Riyadi, Doctoral student at Department of Naval Architecture Institut Teknologi Sepuluh Nopember (ITS). He is responsible for this reseach. His previous experience includes reduce of ship resistace with foil at position in back of ship and investigation the operational mode of fleet in order achieve optimum fuel consumption.

I K A P Utama, a profesor & head of hydrodinamic laboratory at department of naval architecture Institut Teknologi Sepuluh Nopember (ITS), Fellow Member and Vice President for the Asia Region of Royal Institution of Naval Architecture RINA. He is responsible for research in ship resistance, renewable energy, and roughness. His previous experience includes reducing Ship Emissions, etc

W D Aryawan, holds the current position Head of Department at Department of Naval Architecture Institut Teknologi Sepuluh Nopember (ITS). He is responsible for manage of department. His previous experience includes resistance component in symetrical catamaran, etc.

R Rulaningtyas, is a senior lecturer in Physics Department, Faculty of Science and Technology, Universitas Airlangga (UNAIR). Her research interests artificial intelligence, pattern recognition, image processing, and signal processing.

G A Thomas, is the BMT Chair of Maritime Engineering at University College London (UCL). A naval architect, he has over 28 years of research and development experience in the fields of hydrodynamic and fluid-structure interaction working with large range of industry patners. He is currently the Technical Lead for EU project SEDNA: Safe maritime operations under extreme conditions: the Artic case.

EXPERIMENTAL AND NUMERICAL STUDY OF SHIP RESISTANCE DUE TO VARIATION OF HULL VANE® POSITIONING IN THE LONGITUDINAL DIRECTION

I K Suastika, A Firdhaus, R Akbar, W D Aryawan, I K A P Utama, Institut Teknologi Sepuluh Nopember, Indonesia,
S Riyadi, PT. Orela Shipyard, Indonesia, **B Ganapathisubramani,** University of Southampton, UK

SUMMARY

Effects of the Hull Vane® positioning in the longitudinal direction on ship resistance are studied with the aid of computational fluid dynamics (CFD) simulations, verified using data from towing tests. A 31 m hard-chine crew boat was considered with three variations of vane's positioning: vane's leading edge in line with the transom, vane's leading edge one chord length behind the transom and vane's leading edge two chord lengths behind the transom. The vane's submerged depth is three quarter of the ship's draft. The vane's planform is rectangular with NACA 64_1-212 section and aspect ratio of 8.50. At relatively low speed (Fr < 0.5) the Hull Vane® results in an increase of total resistance while at relatively high speed (Fr > 0.5) it results in a decrease of total resistance. The decrease can reach a maximum value of 13.21% for the case of vane's positioning with the leading edge two chord lengths behind the transom. The Hull Vane® was found to work most effectively at the Froude number range between 0.5 and 0.7. At Froude numbers larger than 0.7 the vane's lift becomes too large to result in an excessive bow-down trim, which ultimately results in an increase of total resistance.

NOMENCLATURE

[Symbol]	[Definition] [(Unit)]
α	Vane's angle of attack (° or degree)
ρ	Mass density (kgm^{-3})
Δ	Displacement (t)
μ	Dynamic viscosity (Pa s)
ε, ω	Turbulence dissipation rate (m^2s^{-3})
τ_{ij}	Turbulent shear stress (Nm^{-2})
b	Span length (m)
c	Chord length (m)
c_i	Volume fraction (-)
e	Percent relative error (%)
g	Gravitational acceleration (ms^{-2})
h	Vane's submerged depth (m)
k	Turbulence kinetic energy (m^2s^{-2})
$\mathbf{n}$	Normal vector (-)
p	Pressure field (Nm^{-2})
t	Time (s)
x'/c	Normalized longitudinal coordinate using chord length with the origin at the transom and positive in the rear direction (-)
A	Vane's aspect ratio (-)
B	Boat's breadth (m)
C_D	Drag coefficient (-)
C_L	Lift coefficient (-)
F_1	Blending coefficient (-)
H	Boat's depth (m)
$\mathbf{I}$	Identity vector (-)
N	Number of cells in a simulation (-)
Fr	Froude number (-)
L_{OA}	Length overall (m)
L_{BP}	Length between the perpendiculars (m)
Re_c	Reynolds number based on the chord length (-)
S	Surface (m^2)
T	Boat's draft (m)
$\mathbf{U}$	Velocity field (m/s)
V	Control volume (m^3)
V_s	Ship speed (m/s or kn)
V_{max}	Maximum ship's speed (m/s or kn)

1. INTRODUCTION

Hull Vane® is a fixed vane installed on a ship below the water surface, oriented horizontally in the lateral direction, usually with the vane's leading edge below the transom. The Hull Vane® was invented by van Oossanen in 1992 and patented in 2002. When applied to a ship, it can generate an additional thrust due to the angle between the water inflow and the chord line (affected by the stern form of the vessel), resulting in a net force in the forward direction. Furthermore, it can reduce the bow-up trim, reducing the generation of waves and reducing the vessel's motions in waves, leading to resistance reduction and as such saving of fuel consumption and a higher comfort level for the crew and passengers.

A successful application of the Hull Vane® was reported in [1] when it was applied to a 55 m *MV Karina* fast supply vessel and a 42 m *Alive* superyacht. Furthermore, when a Hull Vane® was applied to a 108 m Holland-class OPV, a reduction in total fuel consumption up to 12.5% was reported in [2] and [3]. At the speed at which most fuel was consumed annually (17.5 knots) the total ship resistance was reduced 15.3%. Other benefits include lower vertical acceleration at the helicopter deck, an increased sailing range and increased top speed.

A better seakeeping performance due to the application of a Hull Vane® was reported in [4] where a ro-pax vessel was considered. The pitch motion and added resistance in waves were reduced due to the application of a Hull Vane®.

The performances of a Hull Vane®, interceptors, trim wedges and ballasting were compared in a study utilizing computational fluid dynamics (CFD) simulations [5]. It was found that the Hull Vane® was the most efficient device in reducing the ship resistance and in improving the seakeeping performance. Considering the AMECRC # 13 high speed displacement vessel, the position of the Hull Vane® was varied in the vertical and longitudinal directions. They reported a reduction in vessel's bow-up trim and sinkage. A resistance reduction was found at Fr between 0.2 and 0.8 reaching a maximum reduction up to 32.4% at Fr = 0.35 for the vane's position with the leading edge approximately 2.5 chord lengths behind the transom. The position of the Hull Vane® in the longitudinal direction was found to have a significant effect on the ship resistance while its position in the vertical direction played a minor role. At Fr between 0.6 and 0.8, the difference in the resulting ship resistance for the different vane's positions was rather small (less than 2%) as the flow behind the transom became more uniform with a resistance reduction between 10 and 12% in this Froude number range.

Considering the positions of the Hull Vane® in vertical and longitudinal directions relative to the ship hull, it is recommended that it should not be placed too close to the hull. On the other hand, it also should not be placed too far below the hull as the angle of attack from the water is decreased. In most application, it should be placed aft of the transom and not to close to the free surface [1]. These are also supported by the results of a study reported in [6] where a crew boat was considered and the vane's submerged elevation was varied with $h/T = 0.75$, 1.0 and 1.5 (h is the vane's submerged depth and T is the ship's draft). From the cases considered, it was found that the most effective vane's position in the vertical direction was $h/T = 0.75$.

Furthermore, considering a crew boat (offshore supply vessel) a resistance reduction of 10% was reported in [7] when a Hull Vane® was installed. The Froude number range in which the Hull Vane® worked most effectively was reported to be between 0.5 and 0.7.

Studies on the Hull Vane® based on CFD simulations and model tests have been reported previously where the device was applied to different types of vessel, such as sailing yachts, motor yachts, supply vessels, container ships, cruise ships and ro-ro vessels. It was found that the Hull Vane® was only suitable in certain applications, not in all cases, since in some applications a resistance increase was found. The buttock angle, the transom submergence and the stern form significantly determine the vane's performance [1]. It also depends on the vessel's speed (most favorable at $0.2 < Fr < 0.7$). As suggested in [1], bulk carriers and crude oil carriers are not suitable for an application of a Hull Vane®. It is more suitable when applied to, for example, supply vessels, ferries or patrol vessels (with length not shorter than 30 m due to investment costs).

Clearly, in retrofitting a Hull Vane® into an existing vessel, each case must be considered separately. This process is very time consuming. A guide line will be very valuable regarding the type of vane's section to be used, the vane's size (for example, percentage of generated vane's lift compared with the ship's displacement) and its positioning in the vertical and longitudinal directions relative to the ship's hull.

The purpose of this study is to investigate effects on the ship resistance of the positioning of the Hull Vane® in the longitudinal direction relative to the ship's hull. Studies on the vane's positioning in the vertical and longitudinal directions were reported in [5] using CFD simulations and in [6] using CFD simulations and towing tests but only considered the vertical positioning. This study pursues that reported in [6] but utilizing different ship. The results can enrich the literature on the applications of a Hull Vane®, particularly, on semi-planing boats.

2. METHODS

A 31 m crew boat with a hard-chine body is considered, designed and built by PT. Orela Shipyard, Indonesia. The ship particulars are summarized in Table 1. The study utilized CFD simulations [8, 9, 10] and towing test experiments.

2.1 VARIATIONS OF VANE POSITIONING

The vane's positioning is varied in the longitudinal direction as illustrated in Figure 1 as follows: (i) vane's leading edge inline with the transom (Case 1), (ii) vane's leading edge one chord length behind the transom (Case 2) and (iii) vane's leading edge two chord lengths behind the transom (Case 3). The vane's submerged depth in all cases is three quarter of the boat's draft [6]. The cases considered are summarized in Table 2, including the reference case of ship without vane (Case 0).

Table 1. Ship particulars.

Parameter	Unit
Length Overall (L_{OA})	31.20 m
Length Between Perpendicular (L_{BP})	28.80 m
Breadth (B)	6.80 m
Depth (H)	2.75 m
Draft (T)	1.40 m
Maximum Speed (V_{max})	27 kn
Displacement (Δ)	104.68 t

Table 2. Summary of the cases considered in this study; *x'* is a longitudinal coordinate with the origin at the transom and positive in the rear direction, *c* is the chord length, *h* is the vane's submerged depth and *T* is the boat's draft.

	x'/c		
	0.0	1.0	2.0
With vane: $h/T = 0.75$	Case 1	Case 2	Case 3
Without vane	Case 0		

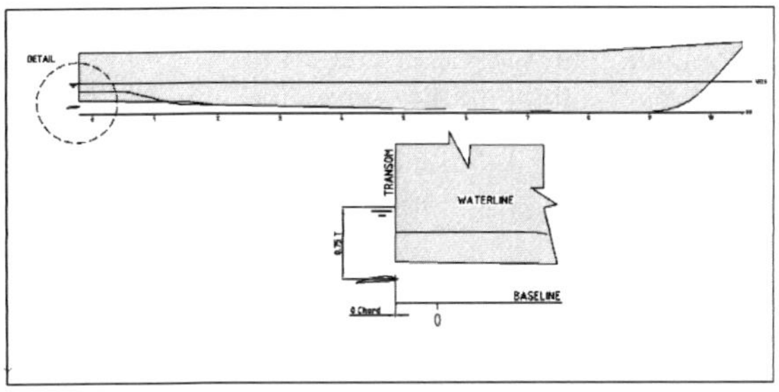

(a) Case 1

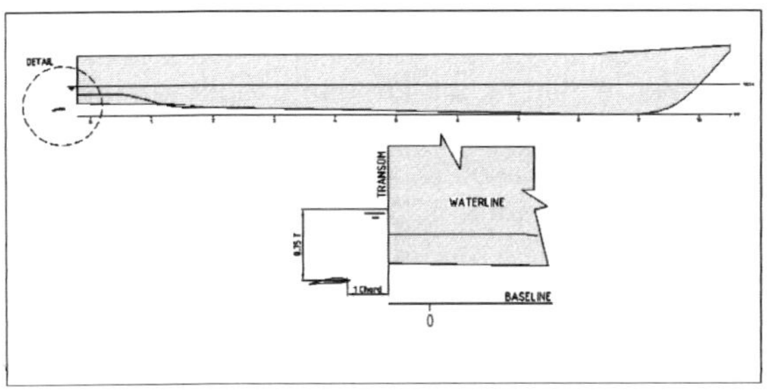

(b) Case 2

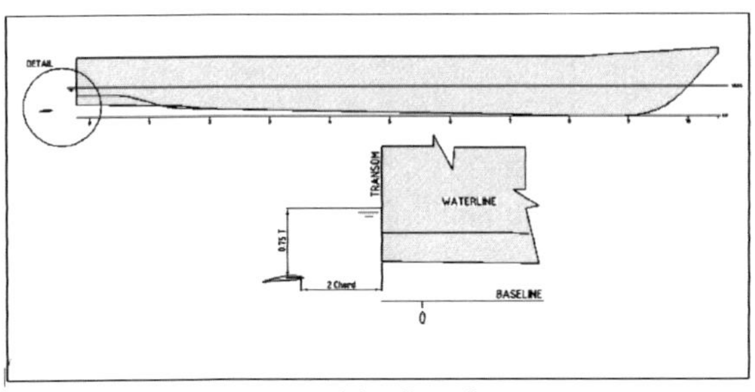

(c) Case 3

Figure 1. Vane's positioning: (a) leading edge inline with the transom (Case 1), (b) leading edge one chord length behind the transom (Case 2) and (c) leading edge two chord lengths behind the transom (Case 3).

2.2 CFD SIMULATIONS

The mass, momentum and volume-fraction conservation equations are represented, respectively, as follows:

$$\frac{\partial}{\partial t}\int_V \rho \mathrm{d}V + \int_S \rho(\mathbf{U}-\mathbf{U}_d)\cdot\mathbf{n}\mathrm{d}S = 0 \quad (1)$$

$$\frac{\partial}{\partial t}\int_V \rho U_i \mathrm{d}V + \int_S \rho U_i(\mathbf{U}-\mathbf{U}_d)\cdot\mathbf{n}\mathrm{d}S = \int_S \left(\tau_{ij}I_j - pI_i\right)\cdot\mathbf{n}\mathrm{d}S + \int_V \rho g_i \,\mathrm{d}V \quad (2)$$

$$\frac{\partial}{\partial t}\int_V c_i \mathrm{d}V + \int_S c_i(\mathbf{U}-\mathbf{U}_d)\cdot\mathbf{n}\mathrm{d}S = 0 \quad (3)$$

In Equations (1), (2) and (3), V is the control volume, bounded by the closed surface S with a normal vector $\mathbf{n}$ directed outward moving at the velocity $\mathbf{U}_d$, $\mathbf{U}$ is the velocity field and p is the pressure field. Furthermore, τ_{ij} is the turbulent (Reynolds) stress tensor, g_i is the component of the gravity vector and I is an identity vector. In Equation (2), the indices $i, j = 1, 2, 3$ but in Equation (3), where two fluids are considered (water and air), the index $i = 1, 2$, where c_i is the volume fraction of fluid i used to distinguish the presence ($c_i = 1$) and the absence ($c_i = 0$) of fluid i. At the interface, $c_i = 1/2$.

The turbulence model used is the SST k-ω model (SST for shear-stress transport), where k is the turbulence kinetic energy and ω is the turbulence specific dissipation rate [11-13]. The main feature of the model is a zonal blending of modelling, using the Wilcox's k-ω model for the flow near solid walls and using the standard k-ε model (transformed into k-ω formulation) for the flow near boundary layer edges and in free-shear layers. The transport equations for k and ω are represented as follows, where the blending coefficient F_1 models the coefficients of the original ω and ε.

$$\frac{\partial(\rho k)}{\partial t} + \frac{\partial}{\partial x_j}\left(\rho U_j k - (\mu + \sigma_k \mu_t)\frac{\partial k}{\partial x_j}\right) = \tau_{t_{ij}} S_{ij} - \beta^* \rho \omega k \quad (4)$$

$$\frac{\partial(\rho \omega)}{\partial t} + \frac{\partial}{\partial x_j}\left(\rho U_j \omega - (\mu + \sigma_\omega \mu_t)\frac{\partial \omega}{\partial x_j}\right) = P_\omega - \beta \rho \omega^2 + 2(1 - F_1)\frac{\rho \sigma_{\omega 2}}{\omega}\frac{\partial k}{\partial x_j}\frac{\partial \omega}{\partial x_j} \quad (5)$$

2.2 (a) Simulations of Vane Alone

Vane-alone simulations were done to obtain the most optimum vane's section and aspect ratio. The lift-drag ratio was used as parameter to determine the optimum vane. Based on previous studies [7], three NACA sections were considered: NACA 4412, NACA 64_1-212 and NACA 21021 [14]. The aspect ratio was kept within the recommended range [15]. For the chosen aspect ratio, the lift-drag ratio was also calculated for varying angle of attack (α). In the calculations, the span was kept constant (the same as the boat's breadth) but the chord was varied.

To obtain the most optimum grid size (number of cells), tests were performed so as the numerical solution fulfills the grid-independence criterion [8]. For each simulation case, which is characterized by the type of vane's section, aspect ratio A and Reynolds number Re_c, the grid independence tests were first carried out.

Figure 2 shows, for example, results of drag calculated using different numbers of cells for the NACA 64_1-212 vane's section, aspect ratio $A = 6.80$ and angle of attack $\alpha = 0$ ($Re_c = 11.7 \times 10^6$). As shown in Figure 2, the drag decreases monotonically with increasing number of cells, which is expected to reach an asymptotic value at very large number of cells.

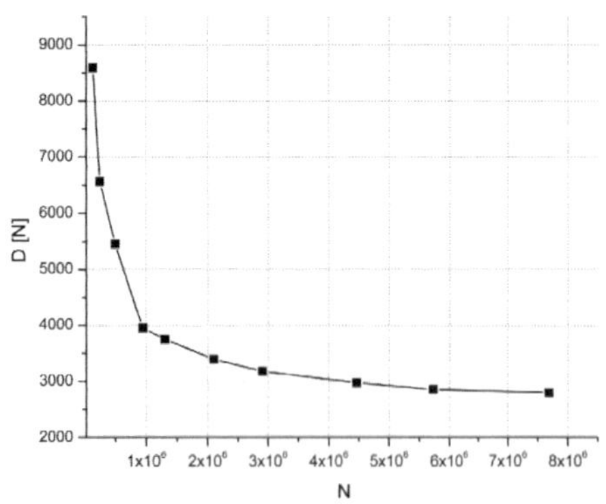

Figure 2. Drag as function of number of cells used in the simulation for the NACA 64_1-212 vane's section and aspect ratio $A = 6.80$ ($Re_c = 11.7 \times 10^6$).

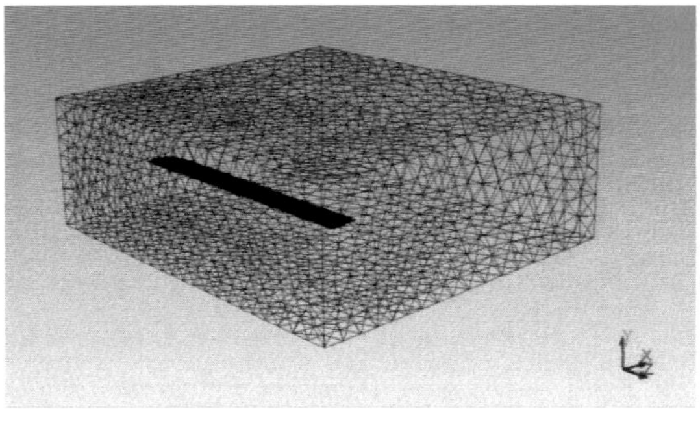

Figure 3. Mesh of the computational domain with the NACA 64_1-212 vane with aspect ratio $A = 8.50$ therein ($Re_c = 9.4 \times 10^6$).

Figure 3 shows, for example, result of meshing the computational domain with the vane therein using number of cells $N = 2.3 \times 10^6$.

The boundary conditions of the computational domain are as follows [9]. The inlet boundary located at 1-c upstream from the leading edge (c is the chord) is defined as a uniform flow with velocity equaling the ship speed (in the simulations the foil is at rest but the water flows). The outlet boundary, at a location 4-c downstream from the trailing edge, is given as that the pressure equals the undisturbed pressure, ensuring no upstream propagation of disturbances. The boundary condition on the foil's surface is defined as no-slip condition. The boundary conditions on the top and bottom walls, at a distance 2-c above and below the foil, respectively, and on the side walls (approximately 7-c aside the model) are defined as free-slip condition.

The root mean square (rms) error criterion with a residual target value of 10^{-5} was used as the criterion for the convergence of the numerical solution.

2.2 (b) Simulations of Ship with and without Vane

The Numeca Fine/Marine® software package [13] was utilized for the CFD simulations, which are based on a finite volume method [9, 10]. The incompressible unsteady ,Reynolds-averaged Navier-Stokes (RANS) equations were solved using the ISIS-CFD code for modelling turbulent multi-phase flows with appropriate boundary conditions [12]. The method of volume of fluid (VOF) was utilized to resolve the free surface boundary [16] for modelling of the generation of waves.

It is common in the application of a numerical method that there is a trade-off between accuracy (which depends on the number of computational cells) and computational cost. To find an optimum number of cells used in the simulations, grid-independence tests were done. Figure 4 shows, for example, the grid independence tests for Case 0. As shown in Figure 4, the ship resistance decreases monotonically with increasing number of cells but has not reached an asymptotic value yet due to limitation of the capacity of the computer. The number of cells of 2.8 $\times 10^6$ was considered as the most optimum number of cells for this case.

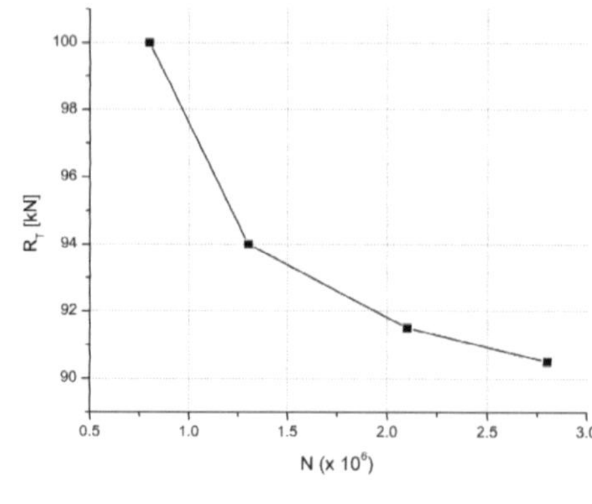

Figure 4. Ship resistance as function of number of cells used in the simulation (Fr = 0.62).

Figure 5 shows a mesh of the vessel with vane attached to the hull by using two struts for Case 3. Because of symmetry, only a half of the vessel was modelled. A non-uniform meshing was utilized: finer grids for the strut (NACA 0010), the vane (NACA 64_1-212) and parts of the hull (bow, stern) where the surface has relatively large curvature.

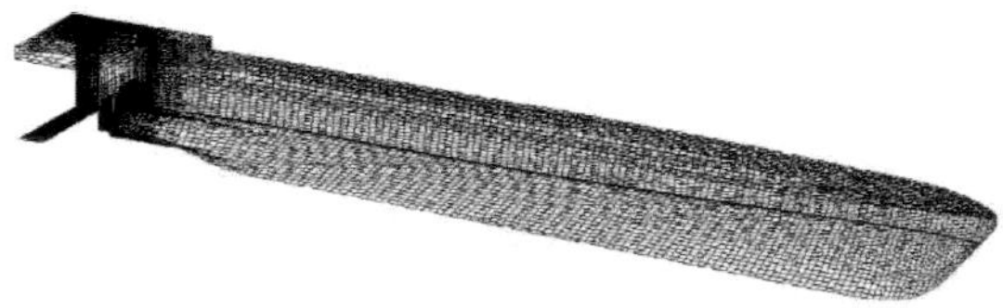

Figure 5. Mesh of half of the vessel, including hull, strut and vane, for Case 3.

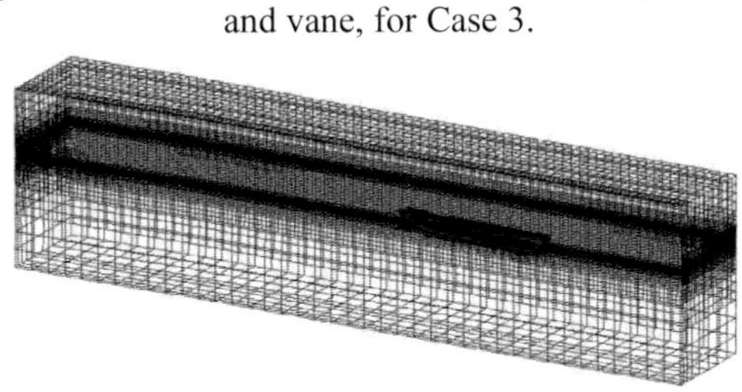

Figure 6: Mesh of the computational domain with the vessel therein for Case 3.

Figure 6 shows the computational domain with the vessel inside it. The inlet of the computational domain is located at 1.0 *L* upstream from the vessel, while the outlet is located at 3.0 *L* behind the vessel. The side wall is located at 1.50 *L* aside the vessel. The bottom wall is located at 1.50 *L* below the vessel and the top wall is located at 0.50 *L* above the vessel (*L* is the length between the perpendiculars). Free surface effects (generation of waves) were modeled in the simulations. The boundary conditions and simulation parameters are summarized in Table 3.

Table 3. Boundary conditions for simulations of ship with and without vane.

Description	Type	Condition
Inlet (Xmin)	EXT	Far field, Vx = 0
Outlet (Xmax)	EXT	Far field, Vx = 0
Bottom (Zmin)	EXT	Update hydrostatic pressure
Top (Zmax)	EXT	Update hydrostatic pressure
Side (Ymin)	MIR	Mirror
Side (Ymax)	EXT	Far field, Vx = 0
Ship hull	SOL	Wall function
Ship deck	SOL	Free slip (zero shear stress)
Motion	Translation in X direction with a given speed	Speed = Ship speed, using one half sinusoidal ramp
	Translation in Z direction; solved motion type	Linear law
	Rotation about the Y axis (pitch); solved motion type	Linear law
Convergence criteria	Order of magnitude of residual decrease	Second order

The root mean square (rms) error criterion with a residual target value of 10^{-5} was used as the criterion for the convergence of the numerical solution.

2.3 TOWING TESTS

To verify the CFD results, towing tests were carried out at the Hydrodynamics Laboratory, Institut Teknologi Sepuluh Nopember, Surabaya, Indonesia. The tank dimensions are as follows: length = 50.0 m, width = 3.0 m and water depth = 2.0 m.

A ship model was designed and manufactured using Froude similarity with a geometrical scale of 1:40. In addition to the ship hull model, models were also made for the vane and the struts. The ship model was made from fiberglass-reinforced plastic (FRP) coated with paint and resin. The vane and the struts were made from brass.

The total resistance was measured by using a load cell. The load cell was connected to a voltage amplifier, which was connected to a computer network in the control room. The load cell was calibrated by using a mass of 0.5 kg before performing a measurement. Six boat speeds were tested: 11, 14, 17, 20, 23 and 26 knots. Figure 7 shows the ship model towed at Fr = 0.62.

Figure 7. Ship model towed at Fr = 0.62. Note the spreading out of water near the bow due to the hard-chine body.

3. RESULTS AND DISCUSSION

3.1 VANE CHARACTERISTICS

The vane characteristics were obtained from CFD simulations of foil alone. To verify the CFD results, these are compared with the theoretical results.

The shifts in lift and drag coefficients due to finite span are given as follows [17]. For a given C_L the horizontal shift in α due to the finite span as compared with the infinite span case is given as:

$$\Delta\alpha = \frac{C_L}{\pi A} \tag{6}$$

Furthermore, for a given α, the increase in C_D due to the finite span as compared with the infinite span case is given as:

$$\Delta C_D = \frac{C_L^2}{\pi A} \tag{7}$$

In Equations (6) and (7), C_D is the drag coefficient, C_L is the lift coefficient, α is the angle of attack and A is the aspect ratio.

Figure 8 shows the lift and drag coefficients for NACA 64_1-212 section with infinite span compared with vane of rectangular planform with NACA 64_1-212 section and aspect ratio A = 8.50 ($Re_c = 1.0 \times 10^6$). The lift curve slope for α between -8° and 12° for the infinite span case is approximately 0.1 per degree and stall takes place at α approximately 14° according to the theoretical prediction [14]. After verification, the horizontal shift of α for a given C_L and the increase of C_D for a given α due to the finite span effects were found to be in accordance with Equations (6) and (7), respectively.

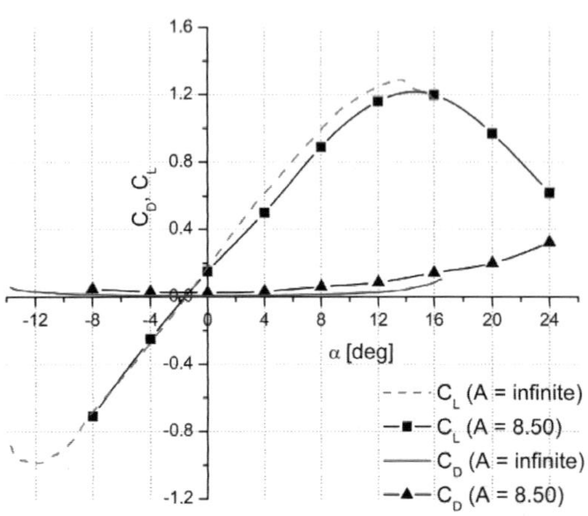

Figure 8: Lift and drag coefficients for NACA 64_1-212 section (theoretical results with infinite aspect ratio A) compared with NACA 64_1-212 vane with aspect ratio of 8.50 obtained from CFD simulations ($Re_c = 1.0 \times 10^6$).

Table 4. Lift-drag ratio for NACA 4412, NACA 21021 and NACA 64_1-212 vanes with varying aspect ratio ($Re_c = 9.4 \times 10^6$). The span b = 6.80 m, which is equal to the boat's breadth. The aspect ratio $A = b/c$, where b is the span length and c is the chord length.

Chord [m]	Aspect Ratio	**Lift-drag Ratio**		
		NACA 4412	**NACA 64_1-212**	**NACA 21021**
0.80	8.50	40.08	41.25	29.77
1.00	6.80	34.90	39.83	26.67
1.20	5.67	31.20	36.54	24.10

Table 4 summarizes the lift-drag ratio for the different vanes considered in this study with the freestream speed of 20 knots (Fr = 0.62), representative for the boat's speed. The corresponding Reynolds number is $Re_c = 9.4 \times 10^6$. Based on the results shown in Table 4, the most optimum section is NACA 64_1-212 with chord length c = 0.8 m and span b = 6.8 m, which is equal to the boat's breadth (aspect ratio $A = b/c$ = 8.50). Furthermore, simulation results show that the value of lift at relatively small angle of attacks lies in the range between 30 and 240 kN, which is 3 to 23% of the boat's static displacement (Δ = 104.68 t = 1027 kN).

3.2 SHIP RESISTANCE WITHOUT VANE

Results of ship without vane (Case 0) are presented in this section. In addition to using towing-tank data, the CFD results were also verified using results from Savitsky's model [18].

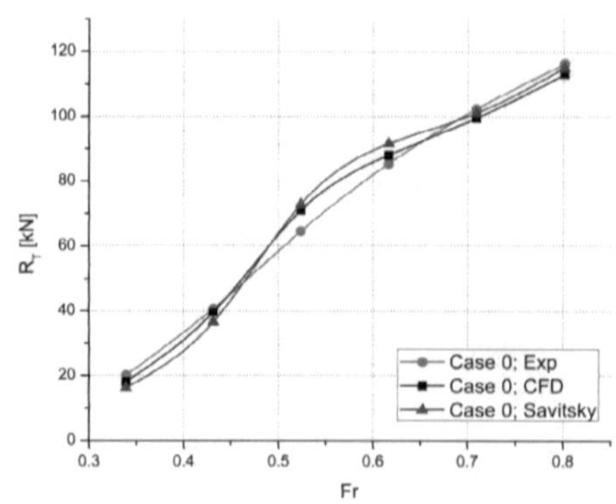

Figure 9: Total resistance as function of Froude number for ship without vane (Case 0) obtained from towing tests, CFD simulations and Savitsky's model [18].

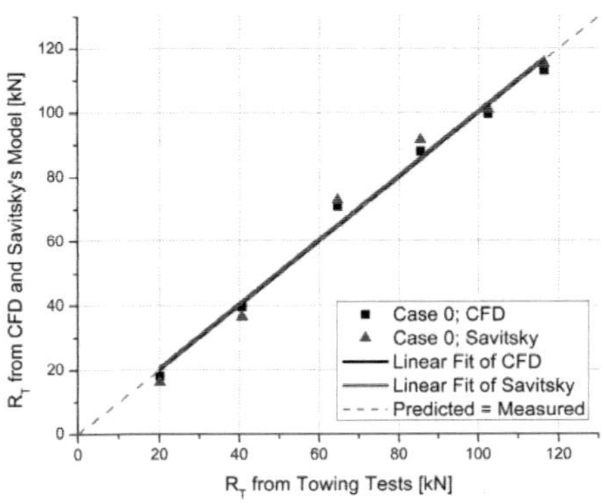

Figure 10. Total resistance obtained from towing tests versus those obtained from CFD and Savitsky's model for Case 0. The coefficient of determination for the linear regression line of measured versus CFD data points is $r^2 = 0.9892$, while that for the measured versus Savitsky's data points is $r^2 = 0.9812$.

Figure 9 shows the total resistance as function of Froude number obtained from towing tests, CFD simulations and Savitsky's model [18]. In addition, Figure 10 shows measured versus CFD resistance data and measured versus Savitsky's resistance data, along with their regression lines compared with the line of a perfect match with a slope of one. As shown in Figure 10 the accuracy of the measurements was acceptable although the agreements between the measured and CFD results and between the measured and Savitsky's results are fair as indicated by the percent relative error shown in Table 5.

Table 5. Percent relative error between measured and CFD results and between measured and Savitsky's results.

Fr	$e_{Exp\text{-}CFD}$ [%]	$e_{Exp\text{-}Sav}$ [%]
0.34	-10.88	-23.93
0.43	-2.79	-11.35
0.52	9.23	11.46
0.62	3.28	6.73
0.71	-2.81	-1.24
0.80	-2.73	-0.99

The resistance plots obtained from the CFD and Savitsky's results (Figure 9) show a hump region in the Froude number range between 0.5 and 0.7. Such a hump region was customarily observed for semi-planing and planing vessels [7, 19].

3.3 SHIP RESISTANCE WITH VANE

Results of resistance for ship with vane are presented in this section. Figure 11 shows the total resistance for the Cases 1, 2 and 3 obtained from the CFD simulations while Figure 12 shows results of total resistance obtained from the towing tests. A comparison between the CFD and towing test results is shown in Figure 13.

Furthermore, Table 6 summarizes the percent relative error between the CFD results and the towing test data while Table 7 summarizes the percent increase of resistance compared with the reference case of ship without vane (Case 0).

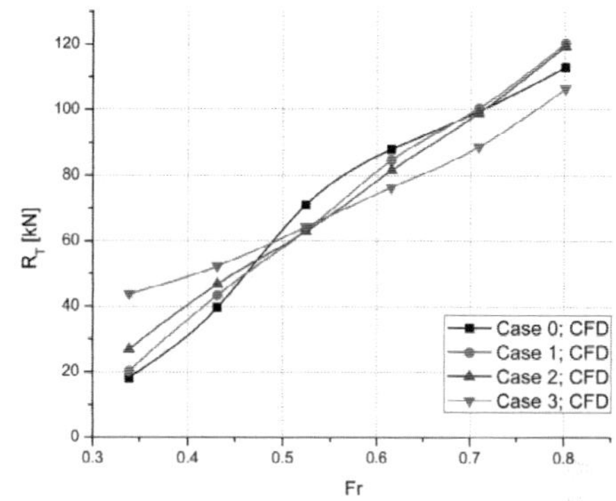

Figure 11. Total resistance as function of Froude number for ship with and without vane obtained from CFD simulations.

Table 6 shows that the percent relative error between the CFD results and the towing test data varies between -6.37 to 6.70. The towing test result for Fr = 0.34 in Case 1 with a percent relative error of -36.72 is an outlier. Although the agreement between the CFD and experimental results is fair, the experimental data favourably verify the CFD results as shown in Figure 13.

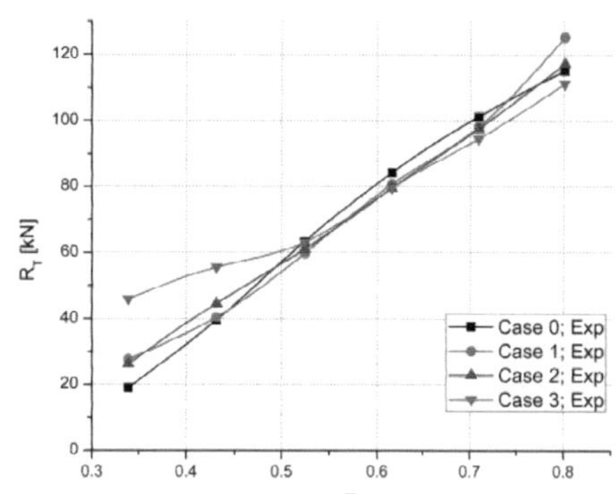

Figure 12. Total resistance as function of Froude number for ship with and without vane obtained from towing tests.

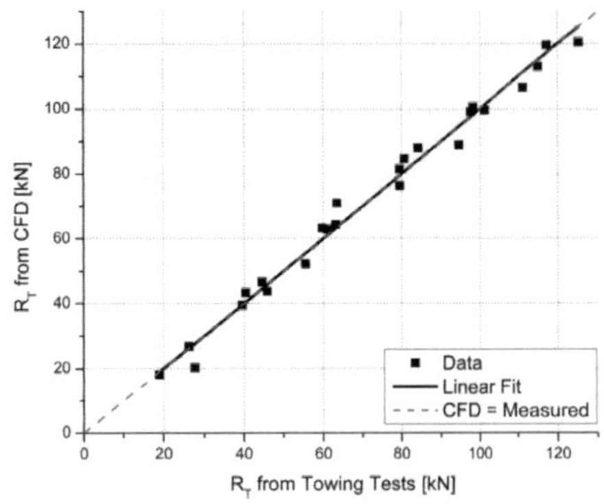

Figure 13. Total resistance obtained from towing tests versus that obtained from CFD for the Cases 0, 1, 2 and 3. The coefficient of determination for the regression line $r^2 = 0.9863$.

At relatively low speed (Fr < 0.5), the Hull Vane® results in an increase of total resistance, as shown in Table 7 and Figures 11 and 12. This can be understood because at relatively low speed, the vane's lift has not been generated optimally and the vane works as an appendage to increase the total ship resistance. However, the increase can reach 141.18% in Case 3 at Fr = 0.34. This increase of ship resistance of 141.18% looks too high, but consistent with the towing test results as shown in Figure 12. Furthermore, for the cases under consideration, the farther the vane's position behind the transom, the larger the increase of ship resistance. These observed phenomena are still under investigation.

At relatively high speed (Fr > 0.5), Case 3 consistently results in a decrease of ship resistance compared with the case without vane (Case 0). The decrease can reach a value of 13.21%. For the Cases 1 and 2, there is a speed range (0.5 < Fr < 0.6 for Case 1 and 0.5 < Fr < 0.7 for Case 2) in which the vane's application results in a decrease of ship resistance. At larger speeds, the application of a vane results in an increase of ship resistance. It is found that the Hull Vane® is most effective at 0.5 < Fr < 0.7. A previous study [20], utilizing round-bilge hull forms, reported a resistance decrease between 8 and 14%, comparable to the present result, when a Hull Vane® was installed at speed in the Froude number range between 0.5 and 0.7.

Table 6. Percent relative error between the CFD results and towing test data. The towing test result for Fr = 0.34 in Case 1 is an outlier.

Fr	Case 1	Case 2	Case 3
0.34	-36.72	2.29	-4.83
0.43	6.70	4.59	-6.37
0.52	5.47	2.78	1.82
0.62	4.72	2.58	-4.14
0.71	2.42	1.52	-6.45
0.80	3.95	1.95	-4.44

Table 7. Percent increase of total resistance compared with the case without vane (Case 0), based on the CFD results.

Fr	Case 1	Case 2	Case 3
0.34	11.65	47.81	141.18
0.43	9.46	18.01	32.00
0.52	-10.94	-11.51	-9.43
0.62	-3.78	-7.32	-13.21
0.71	1.01	-0.53	-10.81
0.80	6.59	5.79	-5.76

In contrast to the results reported in [5] where they found a resistance decrease in the whole Froude number range between 0.2 and 0.8, in this study a resistance increase is found at Fr < 0.5 and a decrease of ship resistance only found in certain Froude number range which depends on the positioning of the vane in the longitudinal direction. The Hull Vane was found most effective at the Froude number range between 0.5 and 0.7 with vane positioning where the leading edge is two chord lengths behind the transom.

A remedy to avoid the resistance increase at relatively low speed (Fr < 0.5) is to release the vane from the water when the boat sails in low speed and put the vane into the water only when the vessel sails at a certain speed range (0.5 < Fr < 0.7).

4. CONCLUSIONS

The effects of vane's positioning in the longitudinal direction were studied by varying the vane's positions where three positions were investigated: vane's leading edge in line with the transom, vane's leading edge one chord length behind the transom and vane's leading edge two chord lengths behind the transom.

At relatively low speed (Fr < 0.5), the Hull Vane® results in an increase of total resistance, which can reach 141.18%. For the cases under consideration, the farther the vane's position behind the transom, the larger the increase of ship resistance.

At relatively high speed (Fr > 0.5), the vane's position with the leading edge two chord lengths behind the transom consistently results in a decrease of ship resistance. The decrease can reach a value of 13.21%. The Hull Vane® most effectively works at the Froude number range between 0.5 and 0.7. At Froude numbers larger than 0.7 the vane's lift becomes too large to result in an excessive bow-down trim, which ultimately results in an increase of total resistance.

5. ACKNOWLEDGEMENTS

This research project was supported by the Ministry of Research, Technology and Higher Education of the Republic of Indonesia under the grant: Penelitian Terapan Unggulan Perguruan Tinggi (PTUPT) with contract no. 5/E1/KP.PTNBH/2019.

6. REFERENCES

1. UITHOF, K., VAN OOSSANEN, P., MOERKE, N., VAN OOSSANEN, P.G., ZAAIJER, K.S., 'An update on the Development of the Hull Vane®', presented at the 9th International Conference on High-Performance Marine Vehicles (HIPER), Athens, 2014.

2. BOUCKAERT, B., UITHOF, K., VAN OOSSANEN, P.G., MOERKE, N., 'Hull Vane® on Hollands-class OPVs-A CFD Analysis of the Effects on Seakeeping', presented at the 13th International Naval Engineering Conference and Exhibition (INEC), Bristol, 2016.

3. BOUCKAERT, B., UITHOF, K., VAN OOSSANEN, P., MOERKE, N., NIENHUIS, B., VAN BERGEN, J., 'A Life-cycle Cost Analysis of the Application of a Hull Vane® to an Offshore Patrol Vessel', Proceeding 13th International Conference on Fast Sea Transport (FAST), Washington D.C., 2015.

4. UITHOF, K., BOUCKAERT, B., VAN OOSANNEN, P.G., MOERKE, N., 'The Effects of the Hull Vane on Ship Motions of Ferries and Ropax Vessels', Design and Operation of Ferries and Ro-Pax Vessels, 25-26 May 2016, London, UK., 2016.

5. UITHOF, K., HAGEMEISTER, N., BOUCKAERT, B., VAN OOSSANEN, P.G., MOERKE, N., 'A Systematic Comparison of the Influence of the Hull Vane®, Interceptor, Trim Wedges, and Ballasting on the Performance of the 50 m AMECRC series #13 Patrol Vessel', Proceeding Warship 2016: Advanced Technology in Naval Design, Construction and Operation, 15-16 June, Bath, UK., 2016.

6. SUASTIKA, K., PRASETYO, B.D., BOAZYUNUS, M., UTAMA, I.K.A.P., RIYADI, S., 'Hull-Vane® Submerged-elevation Optimization for Improved Seakeeping Performance: A Case Study of an Orela Crew Boat', Marine Safety International Conference (Mastic), 9-11 July 2018, Bali, Indonesia, 2018.

7. SUASTIKA, K., HIDAYAT, A., RIYADI, S., 'Effects of the Application of a Stern Foil on Ship Resistance: A Case Study of an Orela Crew Boat', *International Journal of Technology*, 7, pp. 1266-1275, 2017.

8. ANDERSON, J.D, *Computational Fluid Dynamics: The Basic with Applications*, New York: McGraw-Hill, Inc., 1995.

9. VERSTEEG, H.K., MALALASEKERA, W., *An Introduction to Computational Fluid Dynamics: The Finite Volume Method*, Harlow, UK: Longman Scientific, 2007.

10. MOUKALLED, F., MANGANI, L., DARWISH, M., *The Finite Volume Method in Computational Fluid Dynamics*, Switzerland: Springer, 2016.

11. MENTER, F.R., 'Two-equation Eddy Viscosity Turbulence Models for Engineering Applications', *AIAA Journal*, 32(8), pp. 1598-1605, 1994.

12. ISIS-CFD v3.1, 2013. Theoretical Manual, EMN, Ecole Centrale de Nantes, France.

13. FINE/MARINE®, *User Manual: Flow Integrated Enviroment for Marine Hydrodynamics*, Belgium: Numeca International, 2013.

14. ABBOTT, I.H., VON DOENHOFF, A.E., *Theory of Wing Sections (Including a Summary of Airfoil Data)*, New York: Dover Publications, Inc., 1959.

15. VAN WALREE, F., *Computational Methods for Hydrofoil Craft in Steady and Unsteady Flow*, Ph.D thesis, Delft University of Technology, The Netherlands, 1999.

16. HIRT, C. W., NICHOLS, B. D., 'Volume of Fluid (VoF) Method for the Dynamics of Free Boundaries', *Journal of Computational Physics*, 39, pp. 201-225, 1981.

17. WHITE, F.M., *Fluid Mechanics*, 7th ed., New York: McGraw-Hill, 2011.

18. SAVITSKY, D., 'Hydrodynamic Design of Planing Hulls', *Marine Technology*, 1(1), pp. 71–95, 1964.

19. YOUSEFI, R. SHAFAGHAT, R. SHAKERI, M., 'Hydrodynamic Analysis Techniques for High-speed Planing Hulls', *Applied Ocean Research*, 42, pp. 105–113, 2013.

20. ANDREWS, I., AVALA, V.K., SAHOO, P.K., RAMAKRISHNAN, S., 'Resistance Characteristics for High-speed Hull Forms with Vanes', Available at https://www.slideshare.net/karthikavala/final-paper-resistance-characteristics-for-highspeed-hull-forms-with-vanes, Accessed on 30 October 2019.

7. AUTHORS BIOGRAPHY

I K Suastika is associate professor in the field of hydrodynamics at the Department of Naval Architecture, Institut Teknologi Sepuluh Nopember, Surabaya, Indonesia. His research interests include wave-current interactions including wave blocking, boundary layer flows and application of vanes as energy saving devices on single-hull ships and on catamarans.

A Firdhaus is a master student at the Department of Naval Architecture, Institut Teknologi Sepuluh Nopember, Surabaya, Indonesia. He currently studies effects of the application of Hull Vane® on high speed crafts and application of vanes on catamarans as energy saving device.

R Akbar currently conducts his doctoral research at the Department of Naval Architecture, Institut Teknologi Sepuluh Nopember, Surabaya, Indonesia. He currently studies effects of the application of Hull Vane® on high speed crafts and application of vanes on catamarans as energy saving device.

W D Aryawan is a senior lecturer at the Department of Naval Architecture, Institut Teknologi Sepuluh Nopember, Surabaya, Indonesia, where he currently is head of Department. His research interests include ship design, unmanned vehicles and propulsors' performances. He was responsible for the design of the Pusri self-propelled barge, one of Indonesia's most successful design for transporting fertilizer in bulk along the Musi river in South Sumatra.

I K A P Utama is professor of ship hydrodynamics at the Department of Naval Architecture, Institut Teknologi Sepuluh Nopember, Surabaya, Indonesia. He is a fellow and the Vice President for the Asia Region of the Royal Institution of Naval Architects (RINA). He is also a fellow of the Indonesian Academy of Sciences (AIPI). His research interests include ship resistance and propulsion, experimental and computational fluid dynamics, marine renewable energy and maritime education.

S Riyadi is a Technical Director at PT. Orela Shipyard, Indonesia. Currently he conducts a doctoral research at the Department of Naval Architecture, Institut Teknologi Sepuluh Nopember, Surabaya, Indonesia, where he investigates the operational modes of fleets in order to achieve optimum fuel consumption.

B Ganapathisubramani is professor of experimental fluid mechanics in the Aerodynamics and Flight Mechanics Research Group at the University of Southampton, UK. His research interests include understanding the physics and control of turbulent flows in aero-/hydro-dynamic applications, fluid dynamics of biological/bio-inspired systems and development of new experimental and data reduction methods including pressure determination using planar and volumetric velocimetry data, temperature measurements in high-speed flows and innovative ways of examining the data.

NUMERICAL INVESTIGATION INTO THE PRESSURE DISTRIBUTION AND FORM FACTOR EFFECT OF SLENDERBODY CATAMARAN

Sutiyo, E Yuliora, I K A P Utama, Department of Naval Architecture, Institut Teknologi Sepuluh Nopember, Surabaya, Indonesia

SUMMARY

Numerical investigation of pressure distribution between two catamaran demihull was carried out: inside and outside of the configuration. The investigation was conducted using a commercial CFD code (CFX) on the slender model with the ratio of length to width (L/B) is 11.16 and the configuration was made based on different space to length ratio (S/L): 0.2, 0.3, and 0.4 together with different speed of ship. It is apparent that the inside pressure distribution is different from that of outside one, indicating that there is flow interaction between the demihulls. The interaction, which is later expressed as form factor (1+k) values showing that the catamaran form factor is higher than that of the demihull, i.e., 1.42 c.f. 1.26, which demonstrates an increase of approximately 13.6%.

NOMENCLATURE

[Symbol]	[Definition] [(Unit)]
Cv	Coefficient Viscous
v	Velocty (m s^{-1})
L	Length (m)
S	Lateral Spacing (m)
υ	Kinematic viscosity (N s m^{-2})
ρ	Density of water (kg m^{-3})
P	Pressure (N m^{-2})

1. INTRODUCTION

During the last few decades, there has been a significant increase in computer simulations of practical fluid dynamics problems. There are many Computational Fluid Dynamics (CFD) codes, commercial and in-house, that simulate the behaviour of turbulent flows. However, in CFD simulations, it is no longer enough to produce a solution, but also, a credibility analysis of the numerical model should be performed [1].

The use of Reynolds-Averaged Navier Stokes (RANS) methods for the simulation of the water flow around ships has reached a first level of maturity. During the last fifteen years, much progress has been made in the development of robust and accurate computational strategies able to predict flows that contain both viscous and turbulent effects and a free water surface. While this development continues unabated, the application of these methods to full-complexity real-life problems is entering industrial practice.

The current methods can provide a proper evaluation of resistance and wave forces on ships and marine structures, accurate predictions of the flflow fifield useful for improving the ship's hull form design, and local information on the flow enabling the analysis and improvement of appendices and propulsive systems.

The numerical research of hydrodynamics of marine mainly adopts the viscous flow CFD method, which captures more details of the flow field [2]. However, the CFD still needs to be improved for the computational efficiency of viscous flow catamaran hulls interaction. In this present study, the potential flow and pressure method are proposed to rapidly and efficiently achieve the calculation of interaction hull performance. The validation process was also carried out.

RANSE can solve with modern viscous flow code, can simulate the turbulent flow problems efficiently around ships with viscous flow hull interaction. Furthermore, the viscous codes are essential need for investigating some flow characteristics such as flow between and near, where viscous approaches can predict the flow pattern more accurately. Numerical methods are used for solving governing fluid equations with these codes through discretization schemes such as finite differences and finite volume methods.

Viscous resistance represents an integral part of the total resistance of catamaran for intermediate values of Fr, where interference effects are the greatest [3]. Potential flow method has been carried out to determine the lift force single-dead rise hull and catamaran configurations in which hydrodynamic pressure are more pronounced between two catamaran hulls [4]. Therefore, viscous CFD methods should be used for interference investigations rather than potential CFD methods. CFD results show a good agreement with experiments for all separation ratios of catamaran [5].

In this paper, the use of CFD simulation of ship resistance components is presented for the reflex ship model on calm water. The software used for computations was ANSYS-CFX package, which has implemented a RANS solver model. In this paper, CFD simulations of the hull would be conducted with the different pressure distribution. Predicted results for resistance components at various Reynolds numbers in various separation of hull (S/L). The detail information about the geometry of the model, boundary layer, boundary domain, meshing process, study conditions, testing installations would be presented on the following sections of the paper.

2. METHOD

The research methodology that will be used to solve the problem of viscous resistance in the catamaran hull is through numerical simulations using ANSYS CFX. The geometry configuration of the simulated model is the type of displacement of the catamaran with symmetrical hull with some variations in gastric distance transversely. Numerical simulation calculations are intended to get the pressure interaction on the catamaran hull. In more detail, the research methodology diagram is shown in Figure 1

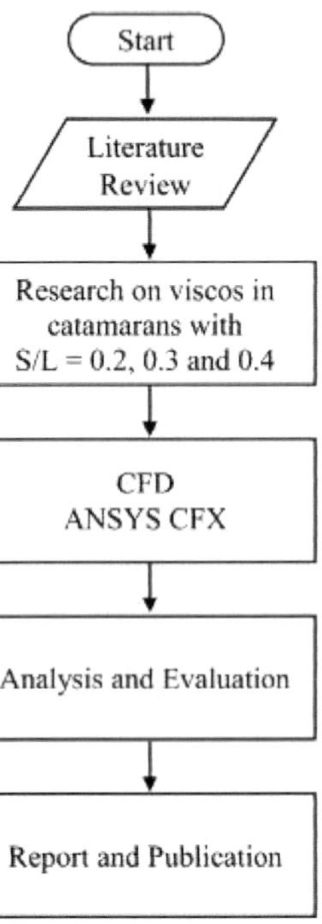

Figure 1. Research Flowchart on Form Factor on Catamaran hull

The computational domain in this study extends for 2L in front of the ship hull, 5L behind the hull, 3L to the side and 2L under the keel of the model. The grid of the RANSE code (ICEM CFD) has been used for building the required hybrid mesh for the code solver. The computational domain of the catamaran reflex model as shown at Figures. 2. And Catamaran ships were analyzed with variations in S / L = 0.2, 0.3 and 0.4 as shown in Figure 3.

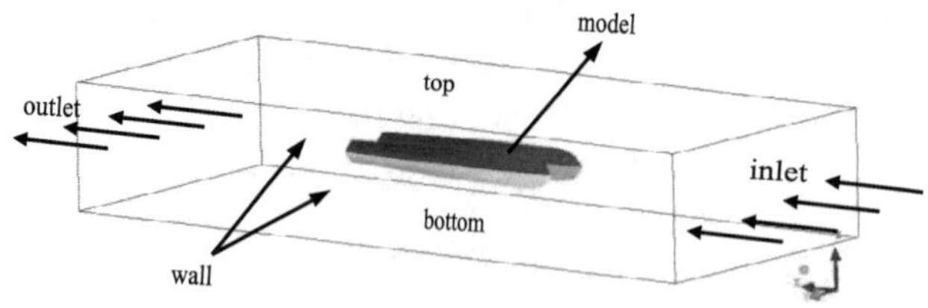

Figure 2. Setting Boundary Condition

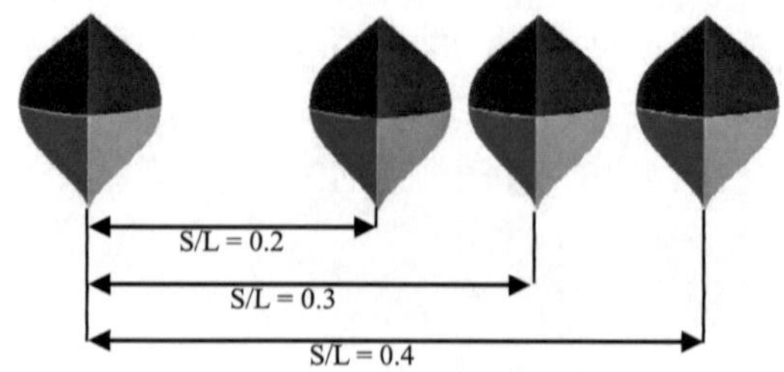

Figure 3. Catamaran with various configuration

The demihull model is designed to have a waterline length of 457.26 mm. The principal particulars of the test model are shown in Table 1.

Table 1. Principal particulars of the model

Parameter	Dimension	Demihull	Unit
Length	L_{WL}	457.26	mm
Breadth	B	40.99	mm
Hight	H	51.99	mm
Separation (S/L =0.2)	$S_{S/L=0.2}$	91.452	mm
Separation (S/L =0.3)	$S_{S/L=0.3}$	137.178	mm
Separation (S/L =0.4)	$S_{S/L=0.4}$	182.904	mm
Wetted Surface Area	WSA	170572.64	mm^2
Volume	V	4799230.32	mm^3
Displacement	Δ	4.81	kg

The boundary conditions of computation used here are bottom, free, inlet, outlet, ship and wall. In the inlet area, the considered parameter is fluid velocity of air. Inlet boundary condition is specified at the front of domain with Reynolds Number 2.89 x105, 3.47 x105, 4.05 x105 and 4.46 x105. Top and side of the domain is refined as free slip walls. Bottom of the domain defined as no slip wall. A pressure outlet condition is defined at the rear of the domain. Hull is defined as no slip wall condition.

The k-ε has given reasonably good results for free-shear-layer flows with relatively small pressure gradients. For wall-bounded flows, the model gives good agreement with experimental results for zero and small mean pressure gradients but is less accurate for large adverse pressure gradients [6]. This has been shown to be better at replicating the flow around the stern of a ship, than simpler models such as k-ε, single and zero equation models [7][8].

The effect of viscosity is taken into account in the Navier-Stokes equations. They are based on three conservation laws; conservation of mass, conservation of momentum, and conservation of energy. The equations are partial differential equations that do not have a defined analytical solution and are therefore solved numerically.

2.1 Fluid Dynamics –Governing equations

Conservation of mass states that the rate of change of mass in an arbitrary material volume is equal to the rate of mass production in that volume.

$$\frac{d}{dt}\int_{V(t)} \rho(x,t)dV = \int_{V(t)} \sigma(x,t)dV \tag{1}$$

In equation (1), $\rho(x,t)$is the density of a particle and $\sigma(x,t)$ is the rate of mass production per volume at time t and position$x$. in the above equation obtain the continuity equation (2)

$$\frac{\partial \rho}{\partial t} + \nabla \rho \boldsymbol{u} = 0 \tag{2}$$

Conservation of momentum states that the rate of change of momentum of a material volume is equal to the total force on the volume. The conservation of momentum law can be written in integral form and using Reynolds transport theorem as presented in equation (3)

$$\int_{V(t)} \rho \frac{Du_i}{Dt} dV = \int_{V(t)} \rho F_i dV + \int_{S(t)} R_i dS \tag{3}$$

Conservation of energy states that the rate of change of energy in a material particle is equal to the amount of energy received by heat and work transferred by the particle. The first law of thermodynamics states (4)

$$\frac{d}{dt}\int_{V(t)} \rho E dV = W + Q \tag{4}$$

where E is the total energy, W is the rate of work done by the surrounding on the fluid and Q is the rate of heat addition

2.2 Turbulent Flow

k-ε is a common and widely used two-equation model. The two transport variables solved for in this model are k, the turbulent kinetic energy and ε, turbulent dissipation. This model was implemented to improve the mixing-length model and proposing turbulent length scales in moderate to complex flows. This model has proven useful for free-shear layer flow although only when the pressure gradient is relatively small [9].

It is assumed for this model that the turbulence viscosity is associated to the turbulence kinetic energy and turbulence dissipation by the expression (5)

$$\mu_t = C_\mu \rho \frac{k^2}{\varepsilon} \tag{5}$$

The transport equations used for solving the transport variables k and ε are as follows: For turbulent kinetic energy k,

$$\frac{\partial(\rho k)}{\partial t} + \frac{\partial}{\partial x_j}(\rho U_j k) = \frac{\partial}{\partial x_j}\left[\left(\mu + \frac{\mu_t}{\sigma_k}\right)\frac{\partial k}{\partial x_j}\right] + P - \rho\varepsilon + P_{kb} \tag{6}$$

For turbulence dissipation ε,

$$\frac{\partial(\rho\varepsilon)}{\partial t} + \frac{\partial}{\partial x_j}(\rho U_j \varepsilon) = \frac{\partial}{\partial x_j}\left[\left(\mu + \frac{\mu_t}{\sigma_\varepsilon}\right)\frac{\partial \varepsilon}{\partial x_j}\right] + \frac{\varepsilon}{k}(C_{\varepsilon 1}P - C_{\varepsilon 2}\rho\varepsilon) \tag{7}$$

In (6) and (7), $P$is the turbulence production due to the viscous forces and is presented by $P=\tau ij(\partial U/i\partial xj)$.. $C\varepsilon 1=1.44$, $C\varepsilon 2=1.92$, $C\mu=0.09$, $\sigma\varepsilon=1.3$ and $\sigma k=1.0$, are the closure coefficients [10].

3. RESULT AND DISCUSSION

The pressure variation around the hull at various demihull spacings is illustrated in Figures 4 to 6. In the case of the catamaran, the pressure and flow velocities that occur on the outer and inner side along the hull show a difference.

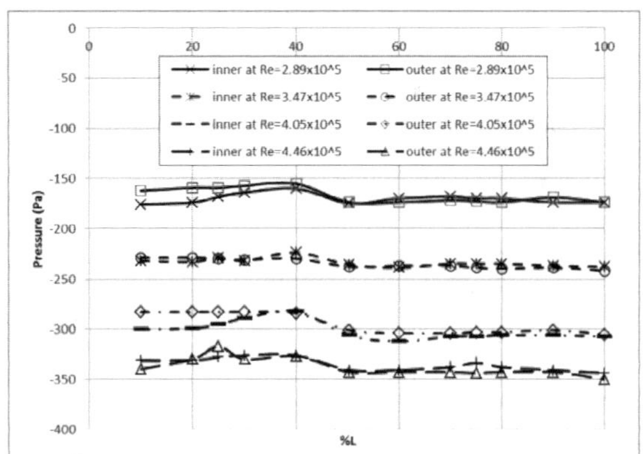

Figure 4. Pressure distribution for catamaran, S/L = 0.2

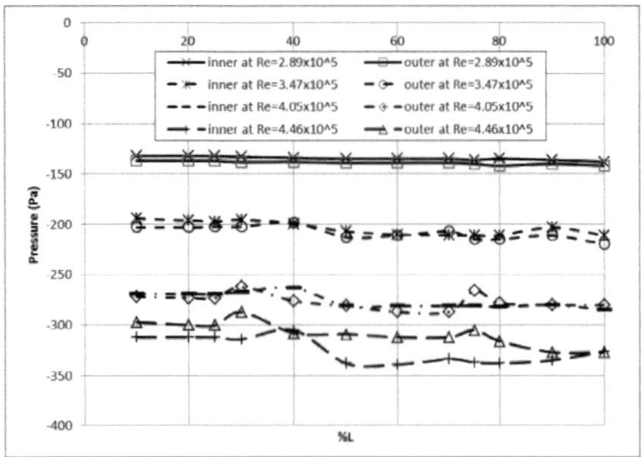

Figure 5. Pressure distribution for catamaran, S/L = 0.3

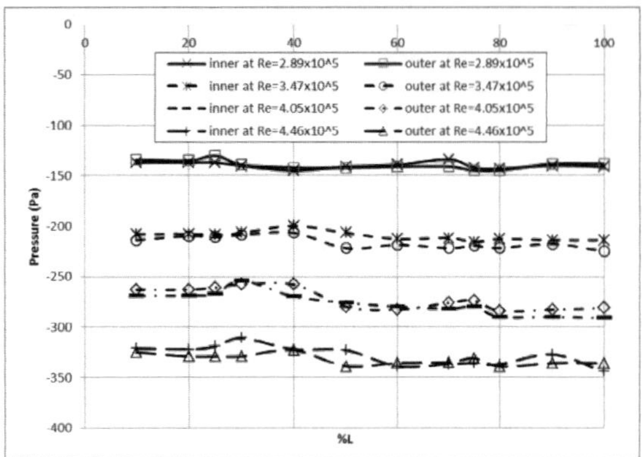

Figure 6. Pressure distribution for catamaran, S/L = 0.4

Table 2 Viscous Resistance Values

Hull Cofiguration	Re	$\int_0^L Cpds$	Cvp'	C_F	C_V = Cvp'+C_F (Corrected)
Demi	2.89 x10^5	0.0845	0.0020	0.0075	0.0095
	3.47 x10^5	0.0803	0.0012	0.0066	0.0078
	4.05 x10^5	0.0770	0.0015	0.0067	0.0083
	4.46 x10^5	0.0744	0.0017	0.0069	0.0086
S/L = 0.2	2.89 x10^5	0.1150	0.0026	0.0062	0.0088
	3.47 x10^5	0.1102	0.0025	0.0059	0.0084
	4.05 x10^5	0.1051	0.0025	0.0057	0.0082
	4.46 x10^5	0.1025	0.0024	0.0056	0.0081
S/L=0.3	2.89 x10^5	0.1091	0.0027	0.0063	0.0089
	3.47 x10^5	0.1060	0.0026	0.0060	0.0086
	4.05 x10^5	0.1005	0.0025	0.0058	0.0084
	4.46 x10^5	0.0970	0.0025	0.0058	0.0082
S/L=0.4	2.89 x10^5	0.1063	0.0027	0.0065	0.0092
	3.47 x10^5	0.1008	0.0027	0.0062	0.0089
	4.05 x10^5	0.0966	0.0026	0.0060	0.0086
	4.46 x10^5	0.0931	0.0026	0.0059	0.0085

From Figures 3 to 5, it can be seen that the pressure ratio (inner pressure/outer pressure) is relatively insensitive to *Re* but shows an increase with increase in hull separation *S/L*.

Since C_P is based on maximum cross-sectional area, CSA (whilst C_{VP} is based on wetted surface area, WSA), the results must be multiplied by a factor (CSA/WSA). The corrected $C_{VP'}$ is then:

$$C'_{vp} = \frac{CSA}{WSA} \times C_{vp} \tag{8}$$

where, CSA= 0.0007778 m^2 and WSA= 0.028444 m^2. The results obtained for viscous pressure resistance and total viscous resistance, together with skin friction resistance, are shown in Table 2.

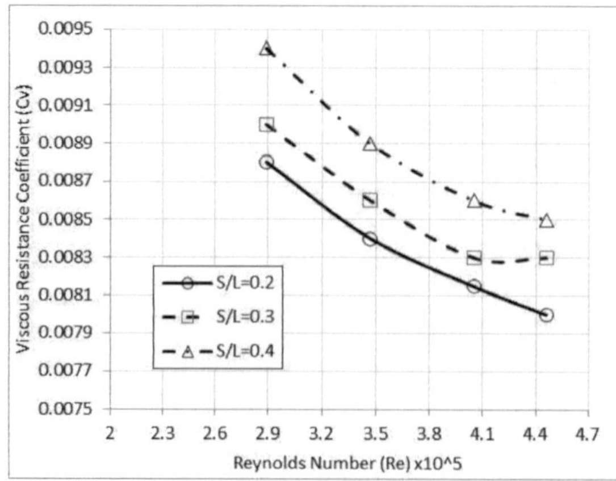

Figure 7. Effect of hull separation on total viscous resistance.

Figure 7 show the effects of *Re* and *S/L* on total viscous resistance C_V. There is a distinct decrease in resistance with increase in *Re*, with the effect decreasing at the highest *Re*.

There is a change in C_V with decreasing hull separation ratio S/L, Fig. 6. It is also describes this phenomenon in his experimental work on the NPL catamaran model [11].

Table 3 shows the test results for the viscous form factor for demihull (monohull) and catamaran, using the highest *Re* value (4.46 x 10^5). The form factor for the demihull (1+*k*) and catamaran (1+*βk*) are derived from C_V/C_F.

Table 3. CFD viscous form factor values

Demihull (1+*k*)	Catamaran (1+*βk*)		
	S/L = 0.2	*S/L* = 0.3	*S/L* = 0.4
1.271	1.475	1.437	1.423

The present study indicated that the ratio of total viscous resistance for the catamaran over that for the demihull is about 1.132, and decreased relatively slowly to about 1.147 as the separation-length ratio *S/L* is increased up to 0.4, see Table 3. These values were found to be broadly similar to published data [12], [13] where the value is about 1.103 and decreases to about 1.081 as *S/L* is increased up to 0.47. Utama (1999) had investigated the drag of catamaran reflex hulls and ellipsoids in proximity, with separation/length ratio (*S/L*) = 0.27, 0.37, 0.47 and 0.57, using a low speed wind tunnel.

The results in Table 2 indicate a viscous interaction of the order of 10-12 per cent of the demihull viscous drag; also, there is little effective change in $(1+\beta k)$ with a change in hull separation *S/L*, a characteristic that had been observed elsewhere such as Utama (1999). It should be kept in mind that the values of $(1+\beta k)$ vary with model shape, size and *Re*, Armstrong (2003).

Viscous form factor value effects on the distance between hull (S/L) and indicate that the higher the distance between the hull (S/L) the smaller the value of the viscous form factor. The emergence of interference and interaction between the two catamaran hull causes a pressure difference and velocity of flow between the catamaran's hull.

The inside (inner) between the demihull an increase in flow velocity due to changes in the distance separation between hull. This increase in disturbance causes the flow velocity changes in the structure of the boundary layer. Friction resistance (skin friction) produced will also experience changes and this will cause changes the value of the form factor on the hull of the ship.

Significant changes in pressure between the inner hull and hull outer significant variations occur in the S/L = 0.2 shown in Figure 8. Where it appears that the pressure between the catamaran hulls is lower than outside the hull, this occurs due to the change of higher speed, which is between hull catamarans. Furthermore, the more significant S/L variation, S/L = 0.3, indicates the pressure between the catamaran hulls even though there has been a decrease. In the variation S/L = 0.4 with Re 3.47 $\times 10^5$ and the more significant shows almost no change in pressure between the catamaran hull and outside the catamaran hull. It can be assumed that there is very little interference.

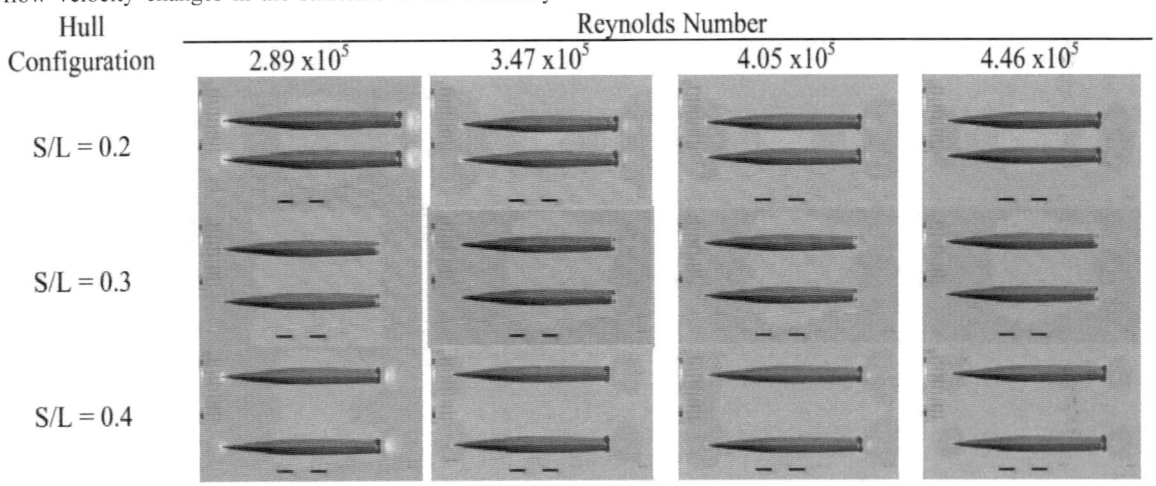

Figure 8. Pressure distribution around hull catamaran

4. CONCLUSION

The effects of various hull separations for a catamaran were analysed in CFD analysis to validate the results. The following conclusions can be drawn :

- Numerical simulation very clearly illustrates the change in flow at inner hulls which causes the result of viscous resistance at S/L = 0.2 is higher than the others. As also, it applies to the pressure acting on the model catamaran.
- The influence of interference and interaction between two ship hulls causes symmetrical flow of water around the demihull to be asymmetrical due to the high pressure and flow velocity that occurs around the hull is relatively unequal or low to the hull centerline.
- Demihull perturbed flow velocity increased in the area around the inside of the hull that alters the structure of the boundary layer. Physical changes the structure of the hull boundary layer causes increased skin friction then change the viscous resistance.

5. ACKNOWLEDGEMENTS

The authors wished our gratitude for the completion of this paper which is part of the World Class Professor implemented at Institut Teknologi Sepuluh Nopember Surabaya

6. REFERENCES

1. Celik, I. (2008), Procedure for Estimation and Reporting of Uncertainty Due to Discretization in CFD Applications, ASME J. Fluids Eng, 130.
2. Frederick, S.; Wang, Z.Y.; Yang, J.; Sadat-Hosseini, H.; Mousaviraad, M.; Bhushan, S.; Diez,

M.; Yoon, S.H.; Wu, P.C.; Yeon, S.M.; et al. Recent progress in CFD for naval architecture and ocean engineering. J. Hydrodyn. 2015, 27, 1–23.

3. Farkas, A., Degiuli, N., Martić, I,. 2017. Numerical investigation into the interaction of resistance components for a series 60 catamaran. Ocean Engineering Volume 146, 1 December 2017, Pages 151-169

4. Ghazi S. Bari, Konstantin I. Matveev. 2017. Hydrodynamics of single-deadrise hulls and their catamaran configurations. International Journal of Naval Architecture and Ocean Engineering, Volume 9, Issue 3, May 2017, Pages 305-314

5. Srinakaew S., Taunton, D. J., Hudson, D. A.. Numerical Study Of Resistance And Form Factor Of High-Speed Catamarans. Transactions of the TSME: JRAME 2019, Volume 7(1)

6. Wilcox, D. C.: Turbulence Modeling for CFD. DCW Industries, Inc., 5354 Palm Drive, La Cafiada, Calif., 1993.

7. Larsson L, Stern F, Bertram V. (2003) Benchmarking of Computational Fluid Dynamics for Ship Flows: The Gothenburg 2000 Workshop. Journal of Ship Research 2003;47:63–81(19).
8.
9. Hino T. (2005) CFD Workshop Tokyo 2005. In: The Proceedings of CFD Workshop Tokyo.

10. Bardina, J.E., Huang, P.G., Coakley, T.J. (1997), "Turbulence Modeling Validation, Testing, and Development", NASA Technical Memorandum 110446

11. Ismail B. Celik. 1999. Introductory Turbulence Modeling. West Virginia University Mechanical & Aerospace Engineering Dept. P.O. Box 6106 Morgantown, WV 26506-6106

12. Armstrong, A. The effect of demihull separation on the frictional resistance of catamarans, Proc. of Seventh International Conference on Fast Sea Transportation, FAST'2003, Ischia, Italy, October, 2003

13. Utama, I. K. A. P., Investigation of the Viscous Resistance Components of Catamaran Forms, PhD Thesis, Department of Ship Science, University of Southampton, UK, 1999.

14. Molland, A.F. and Utama, I K A P. Experimental and numerical investigations into the drag characteristics of a pair of ellipsoids in close proximity, *Proceedings of the Institution of Mechanical Engineers: Journal of Engineering for the Maritime Environment*, Vol. 216 Part M, 2002. pp. 107-11

7. AUTHORS BIOGRAPHY

Sutiyo
Sutiyo is Staff of Simulation and Modeling Laboratory at Universitas Hang Tuah, Surabaya. He is currently completing a Master's Program at the Department of Naval Architecture, Institut Teknologi Sepuluh Nopember (ITS) Surabaya.

Egi Yuliora
Egi Yuliora is Lecturer at Bengkalis State Polytechnic. She is currently a Ph.D. student at the Department of Naval Architecture, Institut Teknologi Sepuluh Nopember (ITS) Surabaya.

I Ketut Aria Pria Utama
I Ketut Aria Pria Utama is a professor of ship hydrodynamics at the Institut Teknologi Sepuluh Nopember (ITS). He obtained his PhD in 1999 from the University of Southampton. He is a fellow member of the Royal Institution of Naval Architects (RINA) since 2006 and member of the Indonesian Academy of Sciences (AIPI) since 2015. His research interests include ship resistance and propulsion, seakeeping, computational fluid dynamics, renewable energy, and maritime engineering education.

ENVIRONMENTAL CONDITION OF INDONESIA'S PORTS ANCHORING AREA

M. A. Kurniawan, D. R. Mauliani, S. Anggara, M. Sodik and M. Irfan, Research and Development Division
Biro Klasifikasi Indonesia, Indonesia

SUMMARY

The properties of temporary anchor consist of the weight of anchor, the chain diameter, and chain length. The requirement are determined based on the classification society rules by using the prescriptive formula. Principle dimension of vessel is the main input for the prescriptive formula to obtain the equipment number, then the number will lead to mass anchor, chain length, and chain diameter as the result on given table. The latest requirement to obtain the equipment number assumed that the maximum current speed is 2.5 m/s, maximum wind speed is 25 m/s. Since the rules of seagoing steel ship assumed occurred in the severe ocean location than the Indonesian domestic water. This conditions may give substantial different result since the domestic water indicates lower wave height, slower current and slower wind speed. The three parameters of maximum current speed, maximum wind speed, and maximum significant wave height are collected from several zones around port's, those are 5 nautical miles, 10 nautical miles and 15 nautical miles from the shore however several port's has dedicated port's anchorage area. Moreover, the depth of each those zones are also identified. All parameters are analysed and the reference parameters that could be used to applicable on Indonesia's port anchorage zone, 38m maximum water depth, 15 m/s for wind speed and 2.1 m/s for current speed. While the results of the analysis for the significant wave height is the average condition of 1.5 m and the average wave period of 1.9 s (Beaufort Scale 4)

Keywords: *port's anchorage area, water depth, wind speed, current speed, significant wave height.*

1. INTRODUCTION

Indonesia archipelago has hundreds ports, from small to large size port, from local to international port, and from traditional to modern port. The government of Indonesia decided to divide the archipelago on to 7 port hub as central port for collecting goods and even passenger from smaller ports around its. The ships which are operated around port usually lay in the temporary anchor near the port or at the anchorage area determined by ports authority.

The anchoring equipment is an important component in the operation of vessels. It is mandatory to follow the design according to Classification Rules and should be in maintenance by Classification Society. This is because the safety of the anchoring equipment affects the safety of the vessel. According to those condition BKI as the classification society of Indonesia conduct some research which is support the adjustment condition and requirement for vessel's operated only in domestic area. Almost 70% of composition registered vessels in BKI are domestic vessels [5]. One among the requirment for adjustment is Temporary mooring equipment consist of anchor weight, chain lenght, and chain diameter. There are also some report from seaman that the length of chain in the locker is much longer than the need for actual condition for anchor release.

This work will investigate 16 location of port in the Indonesian territory. The three parameters of maximum current speed, maximum wind speed, and maximum significant wave height are collected from several zones, those are 5 nautical miles, 10 nautical miles and 15 nautical miles from the shore and the anchorage area of each chosen port. Moreover, the depth of each those zones are also identified. All parameters are analysed and the reference parameters that could be used for applicable on Indonesia's port anchorage area will be introduced.

2. INDONESIA'S PORT

The Government of Indonesia [1], decided define 7 ports as hub port to international port. There are Kuala Tanjung/ Belawan port, Tanjung Priok (new) port, Tanjung Perak port, Kijing port, Bitung port, Makassar port, and Sorong port. Those 7 hub port supported by 36 collector port, 24 main port and 67 SSS (triple S) port all those ports connect with the government program, Tol Laut.

Normally every port should have an anchor zone around the port, depend on the port size, the anchor zone usually has a square shape, the port otority will socialize and inform to maritime stakeholder and the maritime map will be updated. The larger the port size, should provide with larger and more anchor zone around. Not only for waiting a loading or unloading process, bunkering, waiting for a drydocking space, but also many of ships might standby there for waiting a charter.

Based on the data of port location in Indonesia, in this study, there are 18 ports selected for analysis considering many vessels that classed by BKI operate in the area as follows:

• Belawan	• Batam	• Samarinda
• Teluk Jakarta	• Balikpapan	• Asahan
• Semarang	• Bitung	• Tegal
• Surabaya	• Ujung Pandang	• Cirebon
• Banjarmasin	• Pekalongan	• Ambon
• Pontianak	• Merak	• Gresik

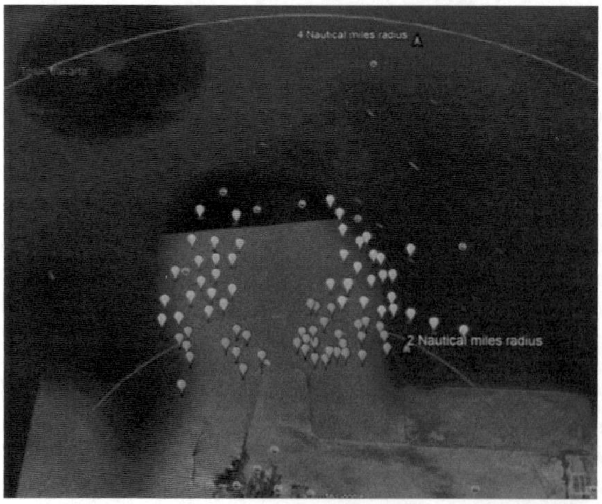

Figure 1. Aerial view around Tanjung Priok Port, Jakarta.

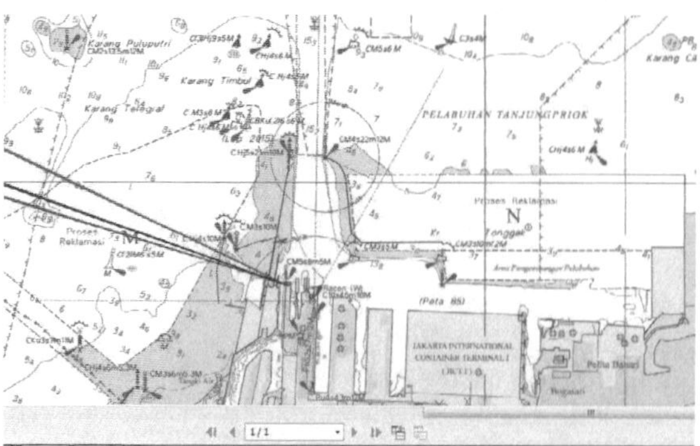

Figure 2. Maritime Map Tanjung Priok Port, Jakarta.

Figure 2 shows the aerial view of Tanjung Priok Port. From the aerial view, we can see that so many of ships are anchoring outside the area of anchoring zone around Tanjung Priok port. The similar condition also happen around others port that are chosen for this study. The seaman lays the anchor under 2 miles radius from nearest port, and many of them also lay the anchor outside 4 miles especially for large and high traffic port like Tanjung Priok and Tanjung Perak. The main reason are the easiness factor to acces the ship when they were in standby station. According to these conditions, this research will take into consideration; i) anchoring zone and ii) the area around the ports based on radius measurement in the analysis.

The analysis area of the chosen ports are divided by 4 (fours) area. That are (a) 5 miles, (b) 10 miles and (c) 15 miles from the coastline and (d) is the anchoring zone. Figure 2 shows the example of maritime map as well as maritime chart of Tanjung Priok port.

All the target areas of chosen ports are investigated the depth of each area. The data of depth water are provided by Goverment agency that has resposible to produce the maritime chart, etc. He is Pusat Hidrografi dan Oseanografi TNI Angkatan Laut. The depth water of all maritime chart of chosen ports are identified for minimum, average and maximum depths. Table 1 shows the identification results of water depth for all chosen ports.

2.1 Anchorage Area

From the maritime charts we can see that the anchorage area are located spesifically on every ports, its depend on the availability of open and free area. The anchoring area should free from underwater cables, include network or telecomunication cables or electricity transmission cables, sea mines area, piping area, coral area, or even in the outside of shipping lanes to access port. In general, a ship not only waiting a loading or unloading process, bunkering, waiting for a drydocking space, but also many of ships will standby there for waiting a charter. She should be anchoring in this anchor zone but several ports still not provided with the anchoring zone.

Anchoring zone should provide the safe location to anchoring, provide enough characteristic of sea bed to obtain maximum grab capacity of anchor. The port authority also makes clustering for anchoring zone based on ship size, local or domestic ship, or international voyage ship. The port authority or port administrator also considers that the port is special zone or free trade zone like Batu Ampar, Batam Port which has direct border with Singapore waterfront. The port authority or administration also consider the anchoring zone should not interfere with the international shipping channel.

The port authority will arrange the technical design for port anchoring zone. The maritime chart/map will propose to the BPI (Indonesian Mapping Body) and DISHIDROS under Indonesian Navy, the authority and the other related body will install marine buoy and marking at the field and held a socialization to maritime stake holder for at least three month probation trial to validate the map/chart and enforced.

2.2 Water depth analysis

Figure 3 shows the example of the maritime chart that representation of Tanjung Emas (Semarang) Port. The area of Tanjung Emas port are cut and the 4 (four) area of water depth analysis is ploted in the chart. The refrences points of each port are decided and each area have diameter from each defined as describe in above. The minimum, maximum water depth of each area in each port are collected.

It is found that the average water depth is ranging from 27 m to 38 m, as shown in figure 4. Furthermore, these conditions will be used as reference analysis representing the Indonesian port condition. Special note is given to Sorong and Bitung ports that are excluded from the depth distribution analysed because of the depth there have different characteristics compared to other ports.

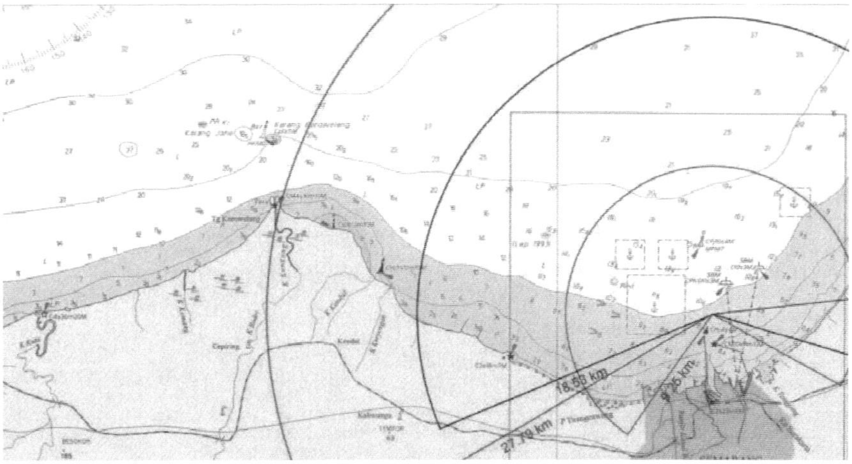

Figure 3. Sample of depth data analysis for Tanjung Emas (Semarang) port

Table 1. Water depth area around the chosen ports (m)

Ports		5 mil			5 mil			10 mil			15 mil			Anchoring Area		
		Shallow	Deep	Avg.	Shallow	Deep	Avg.	Shallow	Deep	Avg.	Shallow	Deep	Avg.	Shallow	Deep	Avg.
BELAWAN	I	-	-	-	7.9	8.8	-	12	23	-	-	30	-	-	18.4	-
BATAM	II	-	-	-	6.3	67	-	34	70	-	-	-	-	26.8	38	-
MERAK JAKARTA	III	-	-	-	4	19.5	-	15	28.5	-	18	37	-	12	18.6	17
SEMARANG	IV	-	-	-	-	20.1	-	-	29	-	-	32	-	-	15.2	13.2
SURABAYA	V	-	-	-	-	7.3	-	-	12.2	-	-	-	-	-	26	-
PONTIANAK	VI	-	-	-	3.5	3.5	-	-	13.3	-	-	21	-	-	17.4	-
BANJARMASIN	VII	-	-	-	4	6	-	1.9	9.2	-	7	23	-	-	-	-
SAMARINDA	VIII	-	-	-	-	5.6	-	-	64	-	-	79	-	-	-	-
BALIKPAPAN	IX	-	-	-	-	6	-	-	35	-	-	52	-	-	27.5	-
UJUNG PANDANG	X	-	-	-	-	26	-	-	29	-	-	43	-	22	33	23
GRESIK	XI	-	-	-	-	5	-	-	10	-	-	15	-	-	-	-
PEKALONGAN	XII	-	-	-	-	25	-	-	36	-	-	41	-	-	-	-
TEGAL	XIII	-	-	-	-	14	-	-	23	-	-	34	-	-	-	-
ASAHAN	XIV	-	-	-	-	4	-	-	39	-	-	48	-	-	-	-
CIREBON	XV	-	-	-	-	9.4	-	-	19.8	-	-	8.8	-	7	10.3	6.4
Ambon	XVI	-	-	24	-	-	-	-	-	-	-	-	-	-	-	-
Manado	XVII	-	-	27	-	-	-	-	-	-	-	-	-	-	-	-
Sorong	XVIII	-	-	27	-	-	-	-	-	-	-	-	-	-	-	-

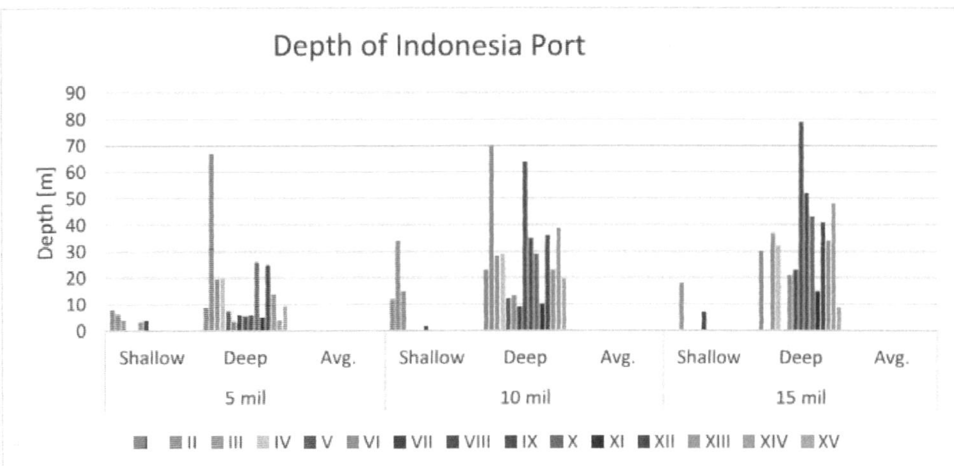

Figure 4. Indonesian port depth conditions

3. ENVIRONMENTAL CONDITION

3.1 References Environmental condition

The anchor's weight and its chain length are determined based on the Classification Society Rules by using the prescriptive formula [2,3]. This prescriptive formula are considered the vessel's principle dimension and assumed that the anchoring equipment used in semi-sheltered or unsheltered anchorages. It is assumed under the following condition:

a) Wind speed 25 m/s, current speed 2.5 m/s, no waves,
   - maximum possible water depth maintaining a scope of six [4],
   - shallow water depth with maximum possible scope

b) Wind speed 11 m/s, current speed 1.54 m/s and significant wave height 2 m, for maximum possible water depth maintaining a scope of six.

It is assumed that those conditions occurred in the severe ocean location, it might be in the North Atlantic Ocean. This conditions may give substantial different if compare with the area that the vessels will be lay-off the anchoring equipment. Those area are specifically may be called as 'port's anchorage area'. Since that the assumed condition of applicable ocean condition of maximum current speed, maximum wind speed and maximum significant wave height are considered for determination of anchoring equipment, it is needed to investigate the special applicable ocean condition based on the port's anchorage area.

3.2 Specific Environmental condition based on port's anchorage area

The strength of the anchor and its chain are influenced by the environmental conditions where the vessel operates. Environmental conditions used to analyse the strength of anchors and its chains are waves, wind and currents.

The wave height and period, wind speed and current speed are determined from data at some point in Indonesia waterways that may represent the condition of the port environment being analysed. There are 10 locations taken to be analysed as samples and shown in Figure 5 and figure 6. These locations have been filtered for the adjacent waters of the islands (closed sea).

From the analysed data is obtained that maximum of maximum wind speed is 15 m/s and maximum of maximum current speed is 2.1 m/s. While the results of the average of significant wave height is 1.5 m and the average wave period is 1.9 second. These conditions is approximately equal to the Scale 4 of Beaufort scale.

These results will be used as a payload to calculate the strength of anchor and its chain. The load for anchor and its chains consists of two loads i.e., static load and dynamic load.

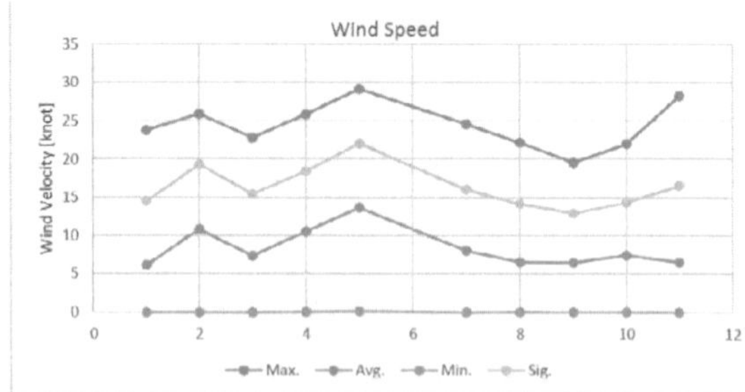

Figure 5. Wind speed in Indonesia waterways

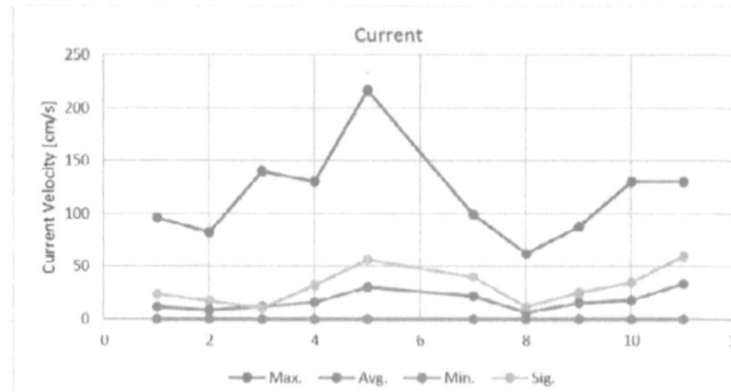

Figure 6. Current speed in Indonesia waterways

The static loads are generated by wind and current acting on the exposed areas of the vessel. These loads are a function of the hull size and shape and the density and velocity of the fluid (air or water) acting on the hull. Current forces are relatively well behaved because current velocity is generally low enough to produce laminar flow around the vessel.

The dynamic loads are caused by the anchor resisting vessel movement and are a function of vessel weight, rate of change in vessel movement (deceleration) and mooring line elasticity. Changes in vessel movement come from two sources; wind gusts and wave action. The loads generated by wind gusts are an order of magnitude less than those caused by wave action. Significant wave height is the average wave height (trough to crest) of the one-third largest waves. Drift force increases with significant wave height and is proportional to wave height squared. Meanwhile, the mean wave period generally result in higher drift forces; when the wave comes into contact with the vessel's hull, the wave is largely reflected.

4. CONCLUSIONS

The 18 (eighteen) ports that located in Indonesian water are chosen in this study. The ports area that located around 5 nmiles, 10 nmiles, 15 nmiles and anchorage area are identified for each chosen ports. The water depth of all chosen ports are identified based on martime charts that supplied by Pushidros TNI AL. The enviromental condition of the average water depth of anchorage area is also analyzed. The results is as follows:

1. The maximum depth to anchorage area in the Indonesian port is 38 m.
2. Environmental conditions used for the analysis of the strength of anchor and its chains in the Indonesian Port is 15 m/s for the wind speed, 2.1 m/s for current speed, and waves with significant height of 1.5 m and an average period of 1.9 s.

5. REFERENCES

1. https://oceanweek.co.id/untuk-efisiensi-7-pelabuhan-terpadu-bisa-saingi-singapura/
2. BKI Rules, Rules for Seagoing Steel Ship, Part I volume 2, BKI (Biro Klasifikasi Indonesia), 2018. Indonesia.
3. UR A1, IACS (International Association Classification Society), Mooring and Anchoring,2017, England.
4. OCIMF (Oil Companies international Marine Forum), Anchoring Systems and Procedures, 2nd Edition, 2010, England.
5. M. A. Kurniawan, Prasetyo F. A, Komariyah, Siti. A Comparison Of Three Different Water Areas And Its Influence For Development Of Rules And Regulation.

THE OPTIMIZATION OF ANCHOR EQUIPMENT DUE TO THE SPECIFIC ANCHORING AREA

D. R. Mauliani, M. Irfan, M. A. Kurniawan, S. Anggara, and M. Sodik, Research and Development Division Biro Klasifikasi Indonesia, Indonesia

SUMMARY

The anchoring equipment are determined considering to the applicable parameters, which are maximum current speed, maximum wind speed, and maximum significant wave height. The requirement is determined based on the classification society rules by using the prescriptive formula. Since the classification society rules assumed occurred in the severe ocean location than the Indonesian domestic water. This condition may give substantial different result to determine the anchoring equipment, since the domestic water indicates lower wave height, slower current and slower wind speed. In the parallel work, the references of those applicable parameters are identified in the Indonesia's port anchorage area. In order to get more reliable result of anchoring equipment, additional work is conducted based on these references applicable parameters of port's anchorage area. This work will analyse the strength of anchor chain from some vessels based on the environmental conditions of the anchorage area. The strength of anchor chain is determined by the safety factor which is used as a reference in determining the New Minimum Breaking Load (MBL). Furthermore, new MBLs are analysed to determine new properties of the anchoring equipment. The new properties are verified by anchor chain strength analysis using the anchor property according to the results that have been generated. Thus, the optimal anchoring equipment will be obtained in accordance with the environmental conditions of its anchorage area.

Keywords: *Anchoring equipment, port's anchorage area, environmental condition, optimization, breaking load*

NOMENCLATURE

L	Length of Chain (m)
OD	Outer Diameter of Chain (m)
Z or EN	Equipment Number
Tp	Peak of Wave Period (s)
D	Depth (m)
SF	Safety Factor
MBL	Minimum Breaking Load (kN)
BKI	Biro Klasifikasi Indonesia
IACS	International Association of Classification Society
UR	Unified Requirements

1. INTRODUCTION

The anchor equipment is determined based on Classification Society Rules with the environmental conditions, those are maximum current speed, maximum wind speed, and maximum significant wave height. It is assumed that those conditions occurred in the severe ocean location, it might be in the North Atlantic Ocean. In the parallel work, the environmental conditions in the port of Indonesia are also identified, so it will be obtained more reliable and optimal results of anchoring equipment.

This work will analyse the strength of anchoring equipment i.e. anchors and its chains using 10 samples of the vessel. The initial stage is identifying the initial anchor chain tension in the Indonesian environment. The next step is analysing the relationship between the anchor chains tension and holding power anchor, so it will be known that the initial tension of the anchor chain obtained does not exceed MBL or proof load anchor.

The strength of the anchor chain is determined by the safety factor which is used as a reference in determining the New Minimum Breaking Load (MBL). Furthermore, new MBLs are analysed to determine new properties of the anchoring equipment. The new properties are verified by anchor chain strength analysis using the anchor property according to the results that have been generated. In addition, the results will be verified using the ratio of the anchor chains length compared to the depth in which the vessel operates. Thus, the optimal anchoring equipment will be obtained in accordance with the environmental conditions of its anchorage area.

2. MOORING ANALYSIS

2.1 Model Definition

The vessel model is chosen with a variation of types and sizes to represent general vessels operating in Indonesia. There are 3 types of vessels in which each consists of 3 sizes. The length of the chain written in Table 1 is the length of the total chain i.e. the summation between the portside and the starboard. Specifically, for barge or pontoon, the length of the anchor chain only requires 50% of the recommended length. The anchor chains length in Table 1 is the length of one anchor chain [3].

2.2 Environmental Condition

The anchor equipment is determined based on Classification Society Rules [3] with the following environmental conditions:

a) Wind speed 25 m/s, current speed 2.5 m/s, no waves, for:
   - maximum possible water depth maintaining a scope of six
   - shallow water depth with maximum possible scope
b) Wind speed 11 m/s, current speed 1.54 m/s and significant wave height 2 m, for maximum possible water depth maintaining a scope of six.

Condition a) is assumed that the anchoring equipment used in sheltered anchorages, whereas for condition b) is assumed in semi-sheltered or unsheltered anchorages.

The environmental conditions for Indonesian port based on the analysis result of the parallel work [4] used for this analysis are as follows:

- The maximum depth to anchorage area in the Indonesian port is 38 m.
- Wind speed 15 m/s, current speed 2.1 m/s, and waves with significant height of 1.5 m and average period of 1.9 s.

2.3 Simulation Method

The strength of the anchor chain will be analysed by referring to the environmental conditions set by the Rules (Case A to D). It aims to know the initial condition. Furthermore, the gap between rules condition and actual condition of Indonesia environment can be known after Case E and F resolved. The gap value will be considered to determine the length reduction of the anchor chain. From the number of load case (See Table 2) and model set, the total simulations performed are 12 x 12 = 144.

Note: *Permanence ratio of the length of the chain will be determined after identification of the entire load case for all completed vessels.*

More clearly related to this method of study can be seen in Figure 1.

Table 1. Model of the analysed vessel

No	Vessel	Type	LBP [m]	Disp. [Ton]	Anchor Chain Length [m]	OD [mm]
1	Vessel A	Tanker	233.0	109424.8	705.3	90
2	Vessel B	Tanker	173.0	39104.2	632.5	68
3	Vessel C	Tanker	149.5	24063.6	577.5	60
4	Vessel D	Pontoon	91.5	8516.0	100.0	12
5	Vessel E	G. Cargo	71.0	2740.3	495.0	35
6	Vessel F	G. Cargo	69.0	2493.0	385.0	34
7	Vessel G	G. Cargo	53.0	-	302.5	22
8	Vessel H	Passenger Ship	52.3	1334.6	385.0	30
9	Vessel I	Passenger Ship	57.4	1338.0	385.0	34
10	Vessel J	Passenger Ship	63.0	1938.0	385.0	36

Table 2. Load Case

Case	Condition	Wave (m)	Current (m/s)	Wind (m/s)	Depth (m)	Release Length (m)		Tp (s)
A	Rules i)	0	2.50	25	27	6D	162	-
B	Rules i)	0	2.50	25	38	6D	228	-
C	Rules ii)	2	1.54	11	27	6D	162	4.078
D	Rules ii)	2	1.54	11	38	6D	228	4.078
E	Actual	1.5	2.10	15	27	6D	162	2.243
F	Actual	1.5	2.10	15	38	6D	228	2.243

Notes:
1. *Rules i) : The environmental condition of Classification Society Rules for condition 2.2 a)*
2. *Rules ii) : The environmental condition of Classification Society Rules for condition 2.2 b)*
3. *Actual : The environmental condition of Indonesian Port*

Figure 1. The Analysis Method of the Optimization of Anchor Equipment Due To the Specific Anchoring Area

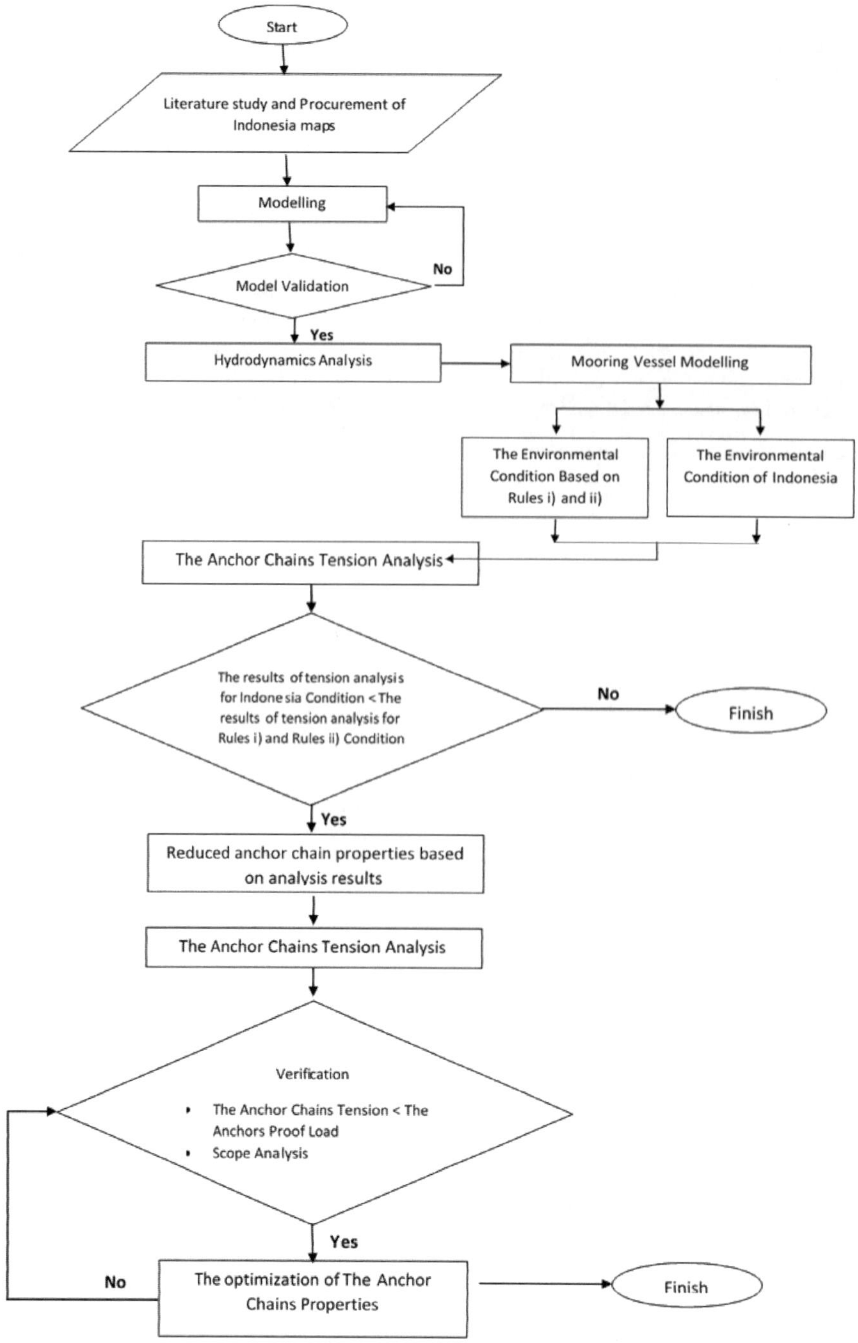

3. ANCHOR EQUIPMENT OPTIMIZATION

3.1 Identify of the Initial Anchor Chain Tension in the Indonesian Environment

The initial stage of study is identifying the condition of the anchor chains tension when operating in Indonesia compared to the environment condition used by Classification Rules as the problem constraints. The modelling conditions of Case A to F (See Table 2) have been applied on 10 vessel models, where the results can be seen in Figure 2. The displayed tension of the anchor chain value is the maximum tension which is reviewed at the fairlead, anchor and touchdown points. From the initial identification of the anchor chains tension can be seen that the average safety factor (ratio of MBL and tension) is obtained by applying environmental conditions for Rules i) and ii) is 6.29. While Indonesia environmental conditions are applied, the average SF is higher up to 8.6. It explained that the Indonesian waterways generates load smaller than the requirement of Classification Society Rules. Therefore, it is possible for reduction of the anchor chain length.

In Figure 2 can be seen the relationship between the maximum tension and MBL of the anchor chain. It can be seen that the maximum tension value of the anchor chain is still relatively far below its MBL.

For the vessel length L ≥ 150 m (Vessel H – J), the analysis results indicate that the condition of the Rules i) generates the anchor chains tension greater than the condition of the Rules ii). However, the condition of the Rules ii) apply wave loads. It can be understood that in principle the mooring system is more influenced by low frequency loads which are winds and currents. Whereas the conditions of the Rules ii) apply wind and current loads smaller than the conditions of the Rules i). On the other side, wave loads have irregular characteristic which can lead an overshooting/significant spike of the anchor chains tension. It is also because the vessels with L ≥ 150 m have wind-exposed areas (wind surface areas) greater than the vessels with L < 150 m (Vessel A – G). This phenomenon may give maximum tension of the anchor chain exceeding averages tension of the anchor chain by the condition of the Rules ii). It also causes the vessels with L ≥ 150 m have greater tension value for the condition of the Rules i) compared to the condition of the Rules ii) and the actual environment conditions. It also applies otherwise to vessels with L < 150 m.

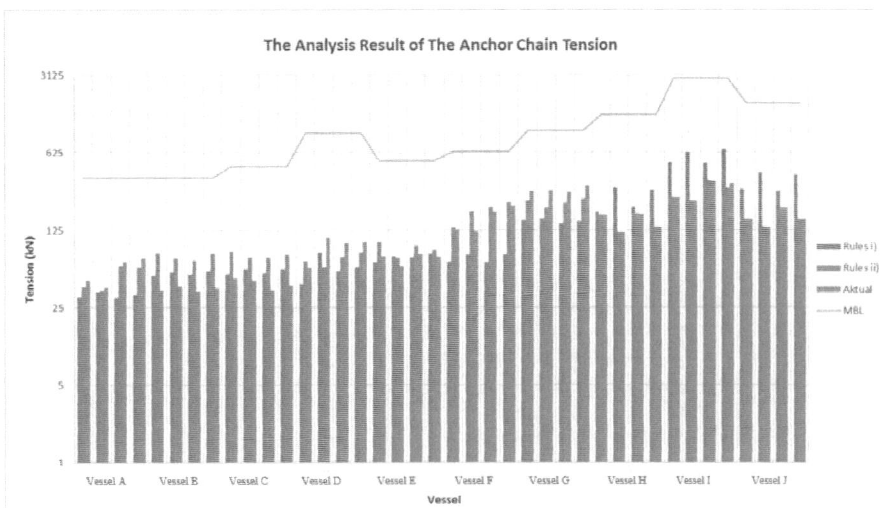

Figure 2. Initial Identification of the Anchor Chain Tension

3.2 The Anchor Chain Tension VS the Holding Power Anchor

Based on IACS research related to anchor equipment in technical background URA1 (Anchoring Equipment), it states that the anchor chains tension is in the holding power anchor range and still under its maximum proof load. It is proved by comparing the tension of the anchor chain obtained using a particular vessel model with a holding power anchor. The ordinary stockless anchor (OSA) type has an efficiency of 1.7 and 3.5, and for the high-holding power (HHP) anchors type has an efficiency of 2.4 and 8. The efficiency is defined as the ratio between the holding power and the anchor weight [1]. Determining the value of efficiency depends on the type of seabed. The lower permeability of the soil has higher the efficiency required. For example, shingle/sand requires higher efficiency compared to the soil of rock and soft mud.

The analysis results of the relationship between the tension of the anchor chain and the holding power anchors are all the tension of the anchor chain obtained still under the anchor load proof (See Figure 3). Most of the tension of the anchor chain from the analysis results indicates that the value is still under the holding power anchor type HHP with efficiency 8 or commonly used for the type of shingle/sand seabed but still above the holding power anchor type HHP with efficiency 2.4. It means that the vessel is not recommended to anchor in seabed condition with soil of soft mud or rock. However, there are several chains of vessels analysed to have tension greater than the holding power anchor type HHP with efficiency 8, but still below the proof load anchor. It cannot be concluded that the anchors are damaged, but it means that the anchors have no grip enough to withstand the tension of the anchor chains.

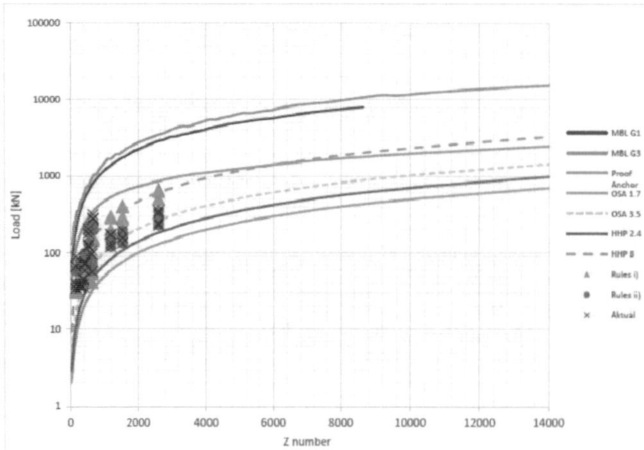

Figure 3. The Relationship between the Anchor Chain Tension and the Holding Power Anchor

3.3 Anchor Chain Properties

3.3 (a) Safety Factor

In this work, the vessel is stated safe if the value of the tension of the anchor chain under the holding power anchor. This level of safety needs to be lowered to a required safety factor which will be used for the reduction of the anchor chain properties (the length and diameter of the anchor chain).

From the analysis results, the tension of the anchor chain is processed to obtain safety factor, which is used to determine the value of new Minimum Breaking Load (MBL). The new MBL is used as a reference for determining the new equipment number (Z). Safety factor of the analysis is processed by dividing the sample of vessels analysed into two parts based on the value of equipment number (Z) (see Figure 4), which are safety factor for $Z < 1000$ and safety factor for $Z \geq 1000$. The determination of the safety factor used as a reference is processed using polynomial approach, thus obtaining the results seen in Figure 5 and 6.

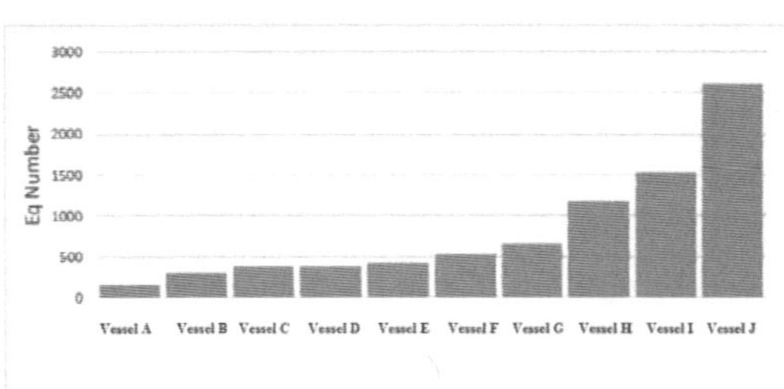

Figure 4. Histogram Equipment Number (Z) Sample

The determination of the reference safety factor for $Z < 1000$ can use the graph in Figure 5 as a reference with different levels of accuracy in each environmental condition. The level of accuracy for the environmental conditions of Rules i) is 100%, the environment conditions of Rules ii) of 99.73%, while for the actual condition of Indonesian waterways by 99.86%. The determination of the anchor chain safety factor to obtain the value of new equipment number in Indonesia waterways for $Z < 1000$ can use the following equation:

$$SF = -1 \cdot 10^{-5}(Z)^2 + 0.0076(Z) + 1.5811$$

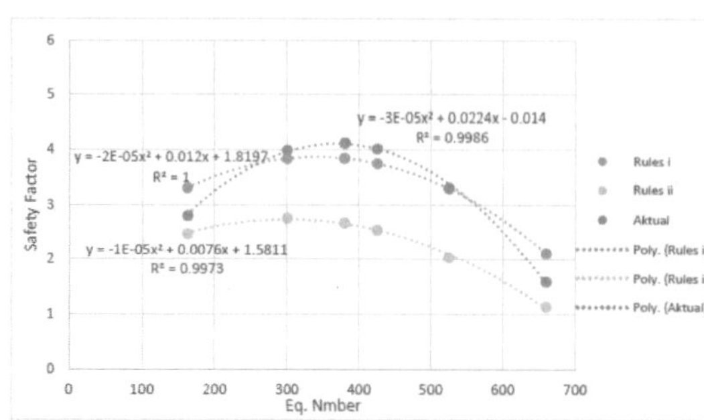

Figure 5. The relationship between Safety Factor and Equipment Number, Z < 1000

Whereas in the determination of the reference safety factor for $Z \geq 1000$ can use the graph in Figure 6 as a reference to the accuracy of each environmental condition has the same value of 100%. The determination of the anchor chain safety factor to obtain the value of new equipment number in Indonesia waterways for Z ≥ 1000 can use the following equation:

$$SF = 2 \bullet 10^{-6}(Z)^2 + 0.0062(Z) + 8.54$$

Where Z is the old equipment number.

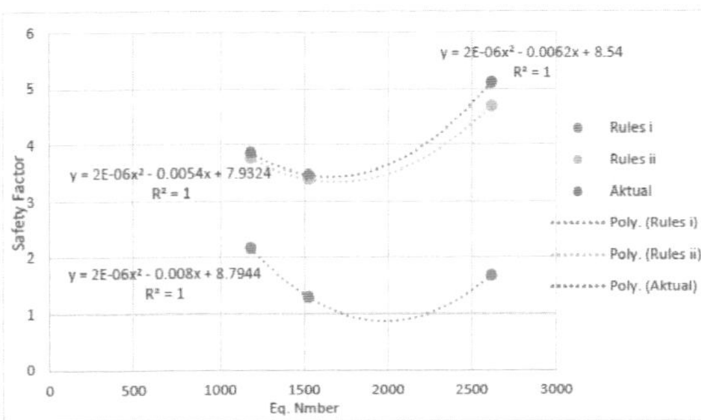

Figure 6. The relationship between Safety Factor and Equipment Number, Z ≥ 1000

From the above graph (Figure 5 & 6) obtained safety factor reference to determine the new equipment number is:

a. Z < 1000, SF = 3
b. Z ≥ 1000, SF = 2.9

3.3 (b) The New Properties of the Anchor Chains

Reduction of the anchor chain properties are obtained by specifying new equipment number. The new equipment number can be determined by comparing the new maximum breaking load obtained from the analysis with the value of the safety factor reference.

The results of the analysis are obtained by the equation that can be used to determine the new anchor chain properties with a level of accuracy for a new anchor chain length of 97.31%, while the level of accuracy for the new anchor chain diameter of 92.84%. (See Figure 7)

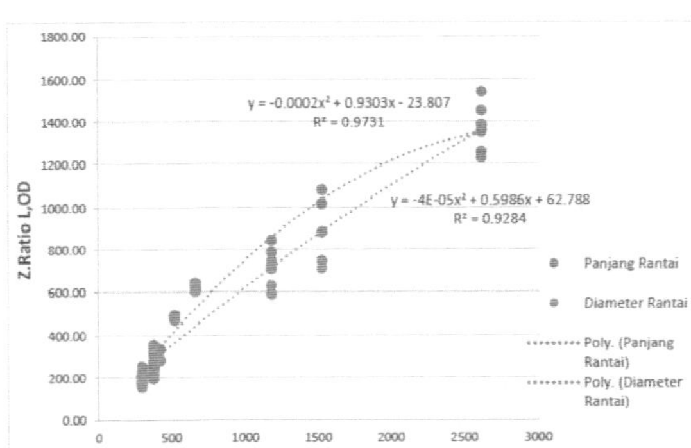

Figure 7. The Relationship between Reducing the Anchor Chain Properties with the Number Equipment

4. VERIFICATION

Verification is necessary to strengthen the results of the analysis, then will be carried out re-analysis with the new anchor chain properties of the analysis results. From the results of repeated analysis, obtained that the maximum safety factor for tension maximum of the anchor chain of 10 samples analyzed model is 2.56 and for the maximum safety factor for proof load anchor is 1.08. It indicates that the reduction of the anchor chain properties is done safely because it still meets the safety factor.

If the value of the anchor chain tension after the reduction of properties corresponding to the analysis results displayed along with the anchor load proof it will look like Figure 8. The anchor chain tension on all simulation models is still under a proof load anchor. The recorded tension of the anchor chain is the maximum value between the fairlead point and anchor point. It seems that the anchor chain tension doesn't change too significantly for models with the Z < 1000, but instead for the Z ≥ 1000. This can be understood because the vessel with a large Z-number has a large mass anyway, so the slight change in the length of the chain will cause significant force change.

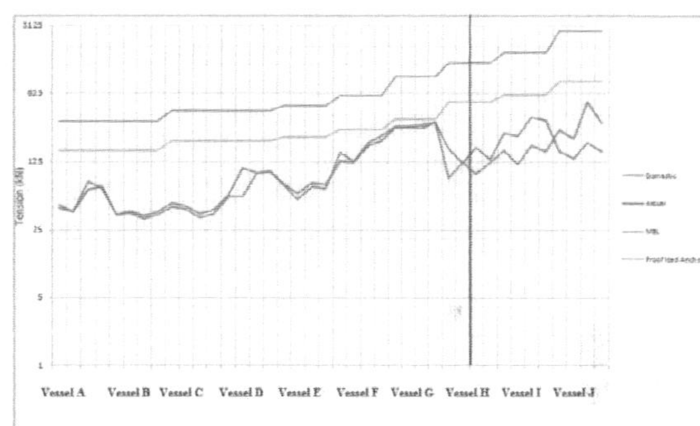

Figure 8. The Changes of the Anchor Chain Tension Due To Reduction of the Anchor Chain Properties

In addition, the verification will also perform scope analysis. Scope in this case is a comparison of the anchor chains length with depth. The process of checking the relationship between scopes of the anchor chains with tension is presented in Figure 9. One of the long scope advantages is to add drag mooring line to support anchor holding power. The shorter the anchor chain will give the effect of increasing tension.

In Figure 9 can be seen that the anchor chains tension has unchanged tendency when the scope is > 4D (D = 38 m), but it will rise significantly when shortened between 2 to 3 of depth. Thus, the determination of the new anchor chains length must also be required that it should not have a scope of not less than 4D.

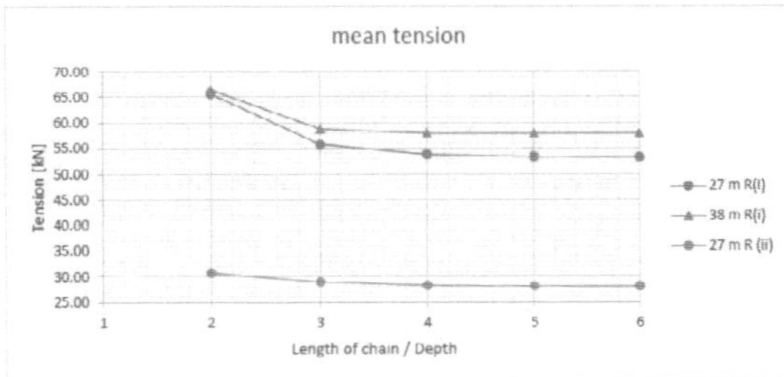

Figure 9. The Scope of the Anchor Chain VS the Tension of the Anchor Chain

5. CONCLUSIONS

From this work the conclusion can be taken as follows:

1. The tension of the anchor chain that have been analysed by applying environmental conditions Indonesia showed lower results compared to the environmental conditions set by the Classification Society Rules, even far below MBL value, so that reduction of the anchor chain properties is possible.
2. Overall the value of the anchor chain tension of the analysis still under the anchor proof load. So this second conclusion supports the previous conclusion.
3. Reduction of the anchor chains can be done with the equations that have been obtained with the note that the vessel is anchored in the Indonesia's port anchorage area with a maximum depth of 38 m.

   The equation to determine the new anchor chain properties are as follows:

   - The New Anchor Chains Length

     $$L(n) = (-0.0002(Z)^2 + 0.903(Z) - 23.807)\, {}^{L}/_{Z} \text{ [m]}$$

   - The New Anchor Chains Diameter

     $$OD(n) = (-4 \cdot 10^{-5}(Z)^2 + 0.599(Z) + 62.79)\, {}^{OD}/_{Z} \text{ [mm]}$$

4. Reduction of the anchor chain properties are already considering safety factor, but does not include corrosion factor.
5. The greater the value of equipment number, the greater the reduction in the anchor chain properties.

6. REFERENCES

1. OCIMF, Anchoring Systems and Procedures, 2n edition 2010. England
2. URA1, IACS, 2017, Mooring and Anchoring. England.
3. BKI Rules, 2019. Rules for Hull (Pt.1, Vol. II), Jakarta, Indonesia: Biro Klasifikasi Indonesia.

[4] M. A. Kurniawan, et.al. 2019. Study on the Environmental Condition of Indonesia's Ports Anchoring Area. Biro Klasifikasi Indonesia, Jakarta

THE HULL GIRDER STRENGTH ANALYSIS DUE TO EQUIPMENT LOAD UNDER LONGITUDINAL BENDING

Muhammad Zubair Muis Alie, Juswan, Wahyuddin Mustafa, Kevin Gabriel Pangalinan and **Nurul Inda Pratiwi,**
Universitas Hasanuddin, Indonesia

SUMMARY

The objective of the present study is to analyze the hull girder strength due to the equipment load of FSO after being converted into FPSO under longitudinal bending. There is no change of the ship's construction; however, much additional equipment after conversion is conducted. One of them is the processing module, where the equipment is placed at the deck part. This additional equipment should be analyzed including their influence to the ultimate strength. The cross-section of FPSO is taken by considering one-frame space. The application of Multi-Point Constraint (MPC) in the numerical method is used. The MPC is placed at the neutral axis position as a reference point on both sides of the cross-section. The cross-section is assumed to remain plane. The midship section is modeled with one frame space. The element type of shell 181 is used on the model. As a simple calculation, the initial imperfections, cracks, and residual welding stress are not taken in the analysis. The ultimate strength obtained by the numerical method is therefore compared with the analytical method and the behavior of the ship in terms of stress distribution and deformation are also presented in this study.

1. INTRODUCTION

The conversion of a ship is now implemented primarily for offshore structure, namely Floating Storage Offloading (FSO) or Floating Production Storage Offloading (FPSO). The purpose of the conversion is to obtain the advantage of the ship payload. Besides, the conversion is also conducted to change the ship function. One of the ship conversion, which is commonly performed, is the ship conversion from FSO to FPSO. Nowadays, about 70% of FPSO is produced from the conversion result. Time-consuming is shorter than a new design that is one of the reasons. Due to this reason, the analysis of the ultimate strength of being converted from FSO to FPSO must be taken into consideration.

The ultimate strength analysis of the ship had been presented by some papers like; The residual strength of an Aframax-class double hull oil tanker damaged in the collision had been assessed by Parunov [1] by considering the influence of the rotation of the neutral axis. The impact of nonlinear finite element method models on the ultimate bending moment for hull girder was studied by Xu [2]. There was two analysis performed; those were implicit static analysis and explicit dynamic analysis. A structural reliability analysis model based on a Bayesian belief network was proposed by Li and Tang [3] for the hull girder collapse risk after accidents. The Bayesian belief network was used to represent random states of variable risk events after accidents, as well as the dependencies between activities, and the structural reliability analysis was used to evaluate the failure probability hull girder for each possible accident conditions. The incidence of collision damage models on an oil tanker and bulk carrier reliability was investigated by Campanile [4] considering the IACS deterministic model against GOLADS/IMO database statistics for collision events, substantiating the probabilistic model. The safety of an oil tanker in intact condition was performed by Campanile [5] to investigate the incidence of load combination methods on hull girder sagging/hogging time-variant failure probability. The simplified approach to the ultimate hull girder strength of asymmetrically damaged ships was conducted by Muis Alie [6] considering the critical element under sagging condition. The residual hull girder strength in intact and damage condition under longitudinal bending moment using nonlinear finite element was conducted by Muis Alie [7], and damages were modelled simply by removing the element on the damaged part. The ultimate hull girder strength considering section modulus under longitudinal bending was analysed by Muis Alie and Latumahina [8] and the cross-section of Ro-Ro ship was taken to be analysed.

In the present study, the analysis of hull girder strength due to the equipment load of FSO after being converted into FPSO under longitudinal bending is conducted. There is no change of the ship's construction, but much additional equipment after conversion. One of them is the processing module, where the equipment is placed at the deck part. For the simple calculation, the one-frame space of FPSO's cross-section is considered. The cross-section is assumed to remain plane. The element type of shell 181 is used on the model. The initial imperfections, cracks, and residual welding stress are not taken into account. The ultimate strength obtained by the numerical method is therefore compared with the analytical method and the behavior of the ship in terms of stress distribution and deformation are also shown in this study.

2. FINITE ELEMENT MODELING

In the present study, the numerical method to analyze the hull girder strength due to load equipment of change function from FSO to FPSO before and after being converted is conducted. The ship has 172 m, 30 m and 18.4 m of length, breadth, and depth of ship, respectively. The midship section consists of two kinds of longitudinal stiffeners those are flat-bar and angle-bar. There is also the inner hull in the cross-section. The element type is shell-181. The shell-181 element applied to all of the cross-section as shown in figure 1.

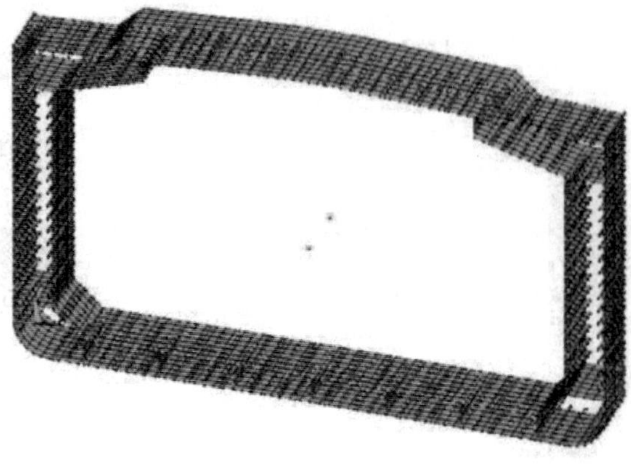

Figure 1. Finite element model of cross section

According to figure 1, there are two points located at the neutral axis. These two points are placed at both sides of the cross-section, and those are used to place the MPC (Multi-Point Constraint) for representing the behavior of the cross-section. The ultimate strength analysis, including the effect of change function from FSO to FPSO, is calculated using the numerical method under sagging condition. The rigidly linked corresponding with the boundary condition where MPC is applied to both sides of the cross-section, as shown in figure 2.

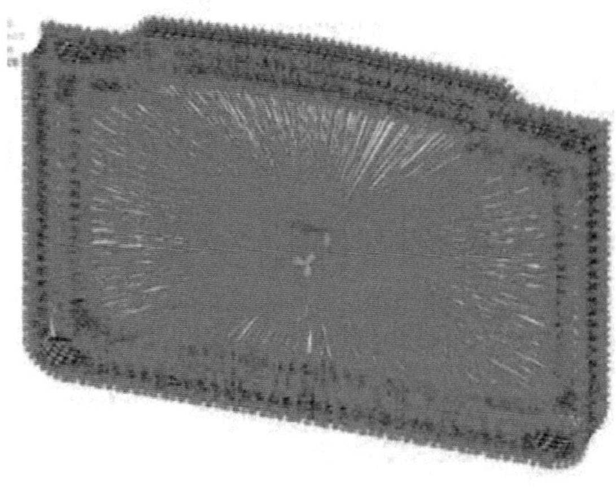

Figure 2. Boundary condition

3. RESULTS AND DISCUSSIONS

The behavior of the ultimate strength analysis is described in terms of working stress distribution. Figures 3 and 4 show the working stress and deformation under the sagging condition. The tension and compression take place at the deck and bottom part since the hull cross-section is under sagging condition.

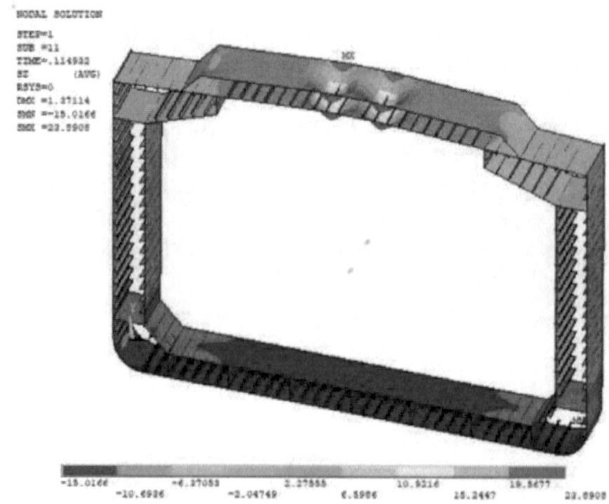

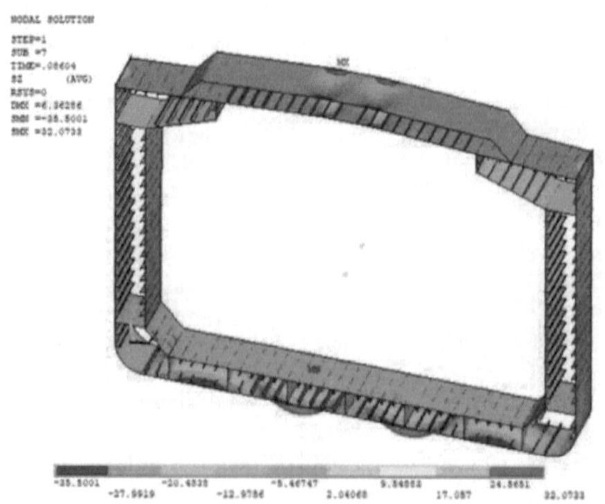

Figure 3. Working stress in sagging

Figure 4. Deformation in sagging

According to figures 3 and 4, the maximum stress and deformation are located in the middle of the cross-section. There are two layers at the deck part of the ship's cross-section. This will be contributed to the ultimate strength since the cross-section is under hogging or sagging condition.

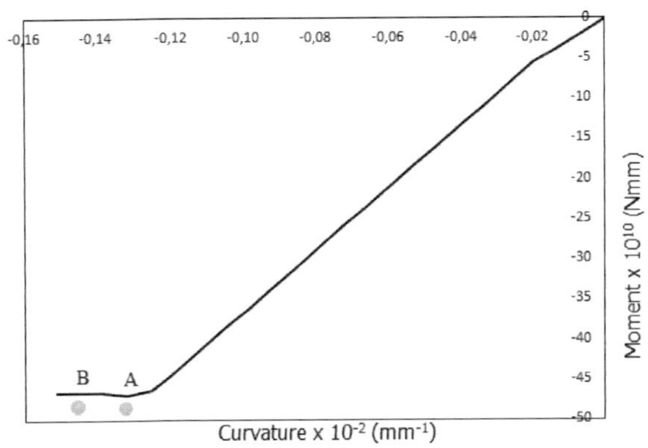

Figure 5. Moment-curvature relationship

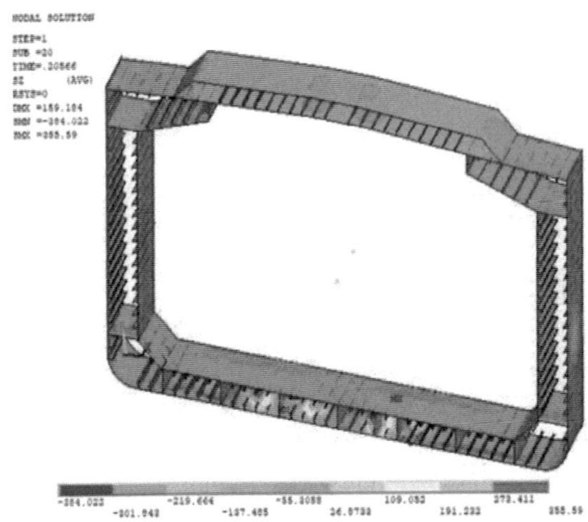

Figure 6. Deformation of FSO at ultimate strength

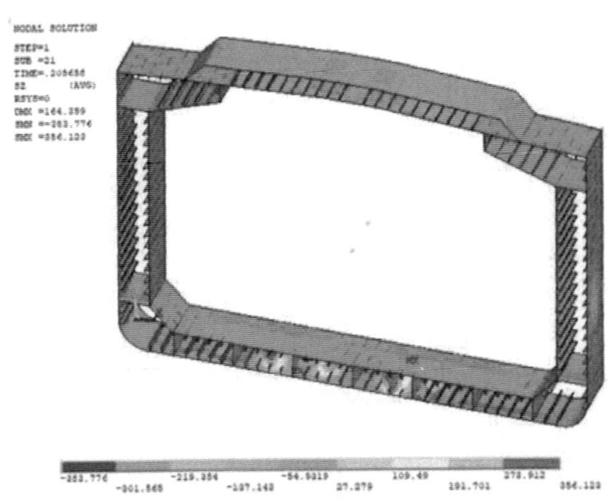

Figure 7. Deformation of FSO at the collapse

Figure 5 shows the moment-curvature relationship of FSO under sagging condition. There are two points at the line, and those point A and B at the ultimate strength and collapse stages. Figures 6 and 7 show the deformation of FSO at point A and B on the ultimate strength and collapse regime, respectively.

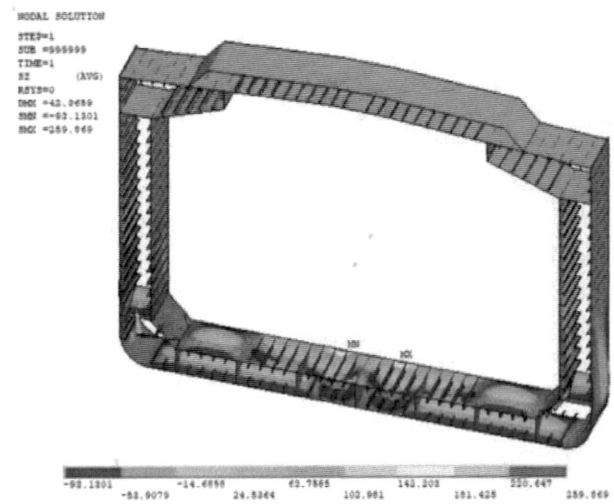

Figure 8. Deformation of FPSO at ultimate strength

According to figures 6 and 7, it is observed that the behavior of deformations is almost similar. It can be seen from figure 5 for the relationship of the moment-curvature curve, where points A and B are almost in a straight line. Therefore, the hull girder behavior of the cross-section at those points is almost similar. Figures 8 and 9 show the deformation of FPSO at ultimate and collapse stages. It is observed that the behavior of the FSO is completely different with FSPO after being converted with additional equipment.

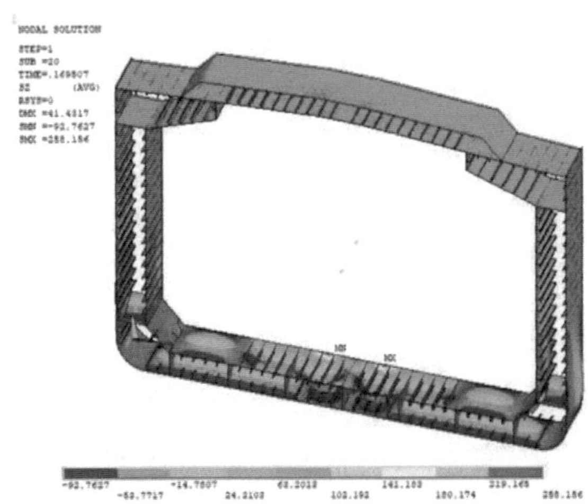

Figure 9. Deformation of FPSO at the collapse

Figure 10 shows the moment-curvature relationship of FPSO under sagging condition. Points A and B are the ultimate strength and collapse regime, respectively.

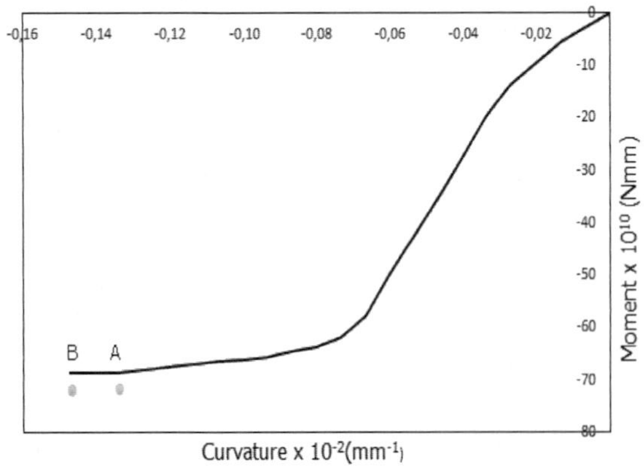

Figure 10. Moment-curvature relationship of FPSO

Figure 11 describes the comparison of the moment-curvature relationship between the analytical method and FEM of FSO under sagging condition. The dashed line represents the result obtained by the analytical method, while a solid line for FEM.

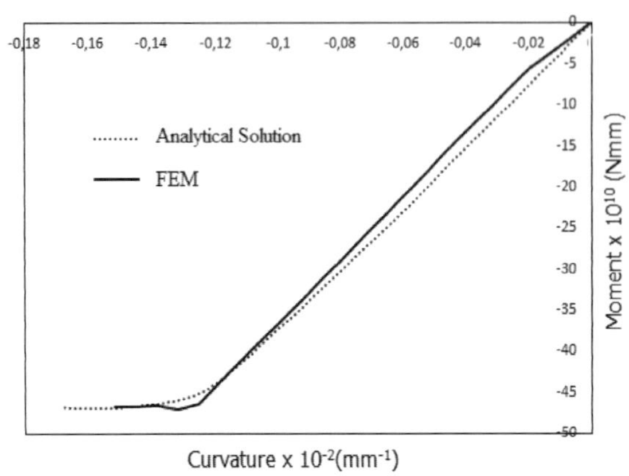

Figure 11. Comparison of moment-curvature FSO

According to figure 11 that the ultimate strength obtained by FEM is a little bit larger than the analytical method. This may be due to the constraint at the cross-section and the load redistribution that is effect to the stress in the elements.

Figure 12 shows the comparison of the ultimate strength obtained by FEM between FSO and FPSO. The solid line represents the ultimate strength of FSO, while FPSO is represented by the dashed line. According to figure 12 that the bending stiffness of FSO and FPSO is different from one another. This behavior is due to the effect of additional equipment and change function where the FSO is converted to FPSO. The effect of the additional equipment which corresponds to loading gives significant influence to the ultimate strength before and after being converted from FSO to FPSO.

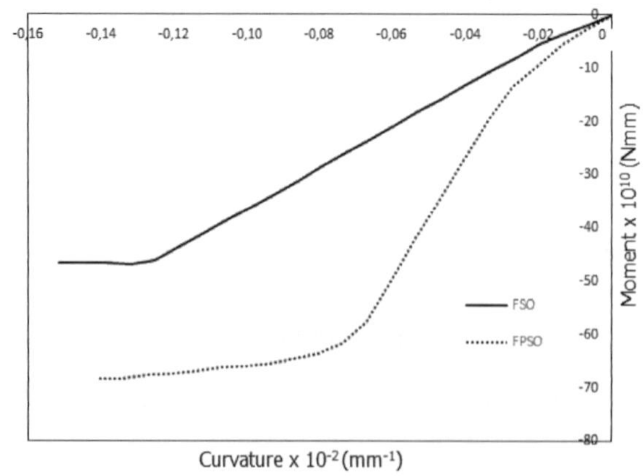

Figure 12. Comparison of moment-curvature between FSO and FPSO

4. CONCLUSIONS

The effect of change function to the ultimate strength from FSO to FPSO is conducted using the nonlinear finite element analysis. The following conclusion is summarized; the impact of change function, namely additional equipment after being converted from FSO to FPSO, is significant to the ultimate strength under sagging condition. Due to additional equipment, the bending stiffness is different between FSO and FPSO. Also, the ultimate strength obtained by nonlinear finite element analysis for FPSO is larger than FSO.

5. REFERENCES

1. Parunov J, Rudan S and Bužančić Primorac B, 'Residual ultimate strength assessment of double hull oil tanker after collision', *Eng. Struct,* 2017.
2. Xu M C, Song Z J and Pan J, ' Study on influence of nonlinear finite element method models on ultimate bending moment for hull girder', *Thin-Walled Struct,* 2017.
3. Li X and Tang W, 'Structural risk analysis model of damaged membrane LNG carriers after grounding based on Bayesian belief networks', *Ocean Eng,* 2019.
4. Campanile A, Piscopo V and Scamardella A, 'Conditional reliability of bulk carriers damaged by ship collisions', *Mar. Struct,* 2018.
5. Campanile A, Piscopo V and Scamardella A, 'Incidence of load combination methods on time-variant oil tanker reliability in intact conditions', *Ocean Eng,* 2017.
6. Muis Alie M Z, 'Simplified approach on the ultimate hull girder strength of asymmetrically

damaged ships', *Int. J. Offshore Polar Eng,* 2018.
7. Muis Alie M Z, Sitepu G, Sade J, Mustafa W, Nugraha A M and Bin Muh. Saleh A, 'Finite Element Analysis on the Hull Girder Ultimate Strength of Asymmetrically Damaged Ships', 2016.
8. Muis Alie M Z and Latumahina S I, 'The Ultimate Hull Girder Strength Analysis Considering Section Modulus Under Longitudinal Bending',2018.

6. AUTHORS BIOGRAPHY

[Muhammad Zubair Muis Alie] Associate Professor at Department of Ocean Engineering, Engineering Faculty, Hasanuddin University. He is the Head of Ocean Structure Analysis Research Laboratory (OSAREL). His previous experience includes the ultimate strength of ship and offshore structures, structural mechanics, finite element analysis and other subjects related to structural strength.

[Juswan] Lecturer at Department of Ocean Engineering, Engineering Faculty, Hasanuddin University. His previous experience includes the planning, design, and construction of the ship and offshore structures.

[Wahyuddin Mustafa] Lecturer at Department of Naval Engineering, Engineering Faculty, Hasanuddin University. He is a member of Ship Construction Laboratory. His previous experience includes the construction and production of the ship and offshore structures.

[Kevin Gabriel Pangalinan] Graduate Student at the Department of Ocean Engineering, Engineering Faculty, Hasanuddin University. He finished his study at Ocean Structure Analysis Research Laboratory (OSAREL). His previous research related to the ultimate strength of Floating Production Storage Offloading (FPSO).

[Nurul Inda Pratiwi] Graduate Student at the Department of Ocean Engineering, Engineering Faculty, Hasanuddin University. She finished her study at Ocean Structure Analysis Research Laboratory (OSAREL). Her previous research related to ultimate strength of Floating Production Storage Offloading (FPSO).

GEOMETRY OPTIMIZATION OF CENTRE BULB TO REDUCE WAVE RESISTANCE ON CATAMARAN SHIP

M Iqbal, E S Hadi, dan **G Pranamya**, Diponegoro University, Indonesia

SUMMARY

A catamaran is a ship with two identical hull forms, which on this type of vessel it has interferences in between its hulls. These interferences cause resistance and a popular topic to research. Multiple factors can cause these interferences, such as the wrong approach to design the hull forms and to set the distance between hulls. On reducing the resistance, there are different approaches that can be used, such as designing streamline demi-hull, adding the bulbous bow, or installing the center bulb. The concept of center bulb itself is considered as a new idea on how to reduce the ship's resistance. Unfortunately, this idea is less explored and needs further research. Thus, this research is trying to use the ellipsoidal shaped center bulb to reduce the wave resistance on a catamaran. The objective is to find the optimum geometry size of length (X1) and diameter (X2) by using Response Surface Methodology (RSM). In this term, the optimum center bulb model can reduce the wave resistance of the catamaran significantly. The result regression formula from Order I is Rw=0.3358X1+0.3969X2+14.5509 and from Order II Rw=-0.061X1+0.096X2+0.373X1^2+0.430X2^2+0.329X1X2. The optimal model of the center bulb can reduce the wave resistance by 11.11% compared to catamaran without an optimized center bulb. Additionally, it can cut down by 5.36% compared to catamaran without a center bulb.

NOMENCLATURE

B	*Breadth (m)*
Cb	*Block coefficient*
Froude Number	*Froude Number*
Lwl	*Length on waterline (m)*
Rt	*Total resistance (N)*
Rv	*Viscosity resistance (N)*
Rw	*Wave resistance (N)*
V	*Speed (m/s)*
S	*Wetted surface area (m^2)*
T	*Draught (m)*

1. INTRODUCTION

A catamaran is a vessel type that has two identical hulls. Those two hulls give the advantage of a spacious deck and make the designer easier to design the interior. On the other hand, it also gives better stability, which provides safety and comfort onboard the ship [1] [2].

With two hulls, the catamaran has interferences between the hulls. This is a popular topic to discuss because interferences can be caused by the wrong approach to design the demi-hull and to set the distance between the hulls. Those two factors have a significant influence on increasing the ship's resistance [3]. It gives a challenge for researchers on how to reduce the resistance caused by interferences.

On how to reduce the resistance, the Luckenby method can be used by modifying the Curved Surface Area (CSA) [4]. By using that method, it can obtain a new hull shape. It successfully reduces the total resistance by 6.5%. Other applications, this approach is useful to increase the seakeeping capability [5].

Another concept to reduce resistance is adding the bulbous bow. This is proven to reduce resistance on a fishing boat [6]. In this case, the bulbous bow used is nabla type, which reduces the total resistance by 10%.

The center bulb concept is considered as the latest idea to reduce the catamaran's resistance. Center bulb is a design component installed in between the hulls [7]. This component also gives better seakeeping ability [8] [9] [10].

The fundamental is; by installing the center bulb, it makes artificial wave interferences. Occurred interferences will break the wave and reduce the wave resistance. Therefore, the total resistance of the catamaran will be decreasing. This concept is visualized in Figure 1.

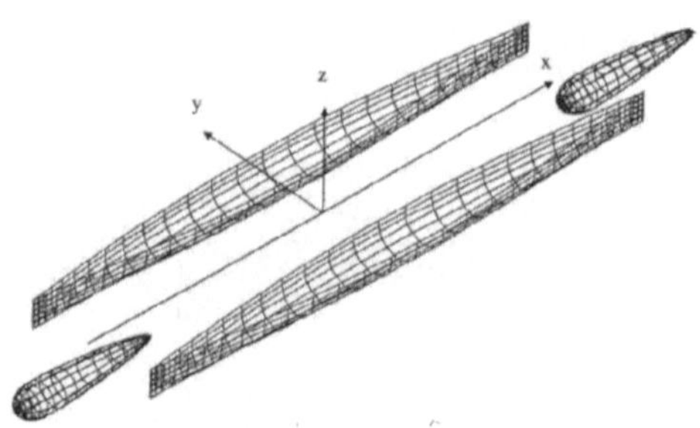

Figure 1. Center bulb concept in a catamaran [11]

Further research on the center bulb is by adding fins to understand more about resistance and seakeeping[8] (Figure 2). The outcome from it, the center bulb can reduce the resistance on Fr>0.3 and increase the resistance on Fr<0.3. In addition, it also improves seakeeping compared to a catamaran without a center bulb. It caused by the descent of the heaving and pitching curve in RAO Curve.

Figure 2. Center bulb concept with fins[8]

In designing the center bulb, it needs a suitable installing location and geometry in order to reduce the resistance. This was the primary objective of Danisman research, which tried to optimized installing location and geometry of the ellipsoidal shaped center bulb by using Artificial Neural Networks (ANN) [7]. This approach can reduce wave resistance by 15% in numeric mode and 13% in the towing tank. The center bulb also installed in a catamaran fishing boat, and it can reduce the resistance by 25.76%.

The application of the center bulb needs further research. Therefore, this research is trying to study the ellipsoidal shaped center bulb based on the assumption that this component can reduce resistance. The goal is; this type of center bulb can make an artificial interference and break the wave in between the hulls of the catamaran.

To obtain the best possible outcome, it needs an optimization method. In this case, the method is used to find the optimum value of length and diameter by using Response Surface Methodology (RSM). The method itself is considered practical and economical and can be used on a variety of variables by evaluating its responses [12]. It gives another advantage that the method does not need a wide variety of Design of Experiments (DoE), and it only takes a small amount of time to experiment [13].

RSM has already been used in many types of research. Such as, to optimize thickness, diameter, and width of a hollow cylinder to produce maximum Specific Energy-Absorbing (SEA) and to minimize Collapse Load (F) [14]. In mechanical engineering, RSM is [15] used to maximize brake power, brake thermal efficiency, minimizing Brake Specific Fuel Consumption (BSFC), and gas emissions by optimizing three variables. Those variables are fuel, speed, and throttle valve.

RSM itself is very rare to use in naval engineering. With that sense, this research is trying to apply this method to reduce resistance in a catamaran. In this research, the optimized variables are length and width from the ellipsoidal shaped center bulb.

The end of this research is; to obtain the optimum length and diameter from the ellipsoidal shaped center bulb by using the RSM method. The definition of the optimum is the geometry center bulb that can reduce wave resistance significantly. Also, this research is trying to find other factors that interfere with wave resistance on a catamaran.

2. METHODOLOGY

2.1. Research Object

The hull form used in this research is from the National Physical Library (NPL) type 4a. The geometry and the Experimental Fluid Dynamics (EFD) result are obtained from Southampton University [16]. To understand more about the main dimension of NPL 4a, look at Figure 3 and Table 1.

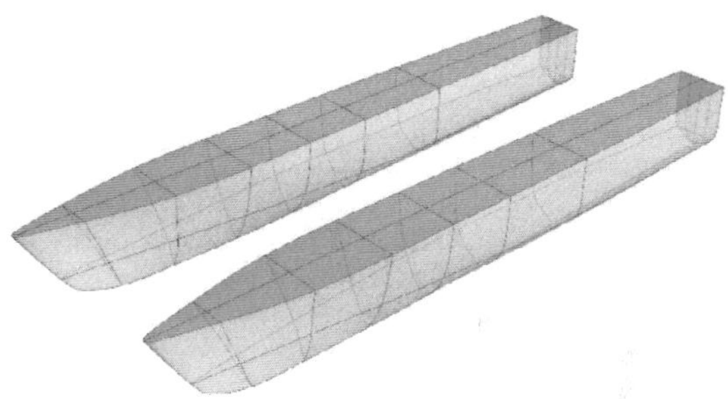

Figure 3. 3D model of NPL 4a

Table 1. Main dimension of NPL 4a

Lwl	1.60 m
B demi-hull	0.15 m
T	0.10 m
Cb	0.395
S	0.346 m^2
S/L	0.3
Displacement	0.0102 ton

Table 2. Total Resistance Coefficient S/L = 0.3

S/L=0.3	
Fr	Ct
0.2	7.145
0.25	7.685
0.3	8.491
0.35	8.653
0.4	9.84
0.45	11.831
0.5	10.948
0.55	9.719
0.6	8.756
0.65	8.003
0.7	7.509

The concept of the center bulb is adopted from Danisman's research. The center bulb is ellipsoidal shaped and used as the initial geometry. The size is obtained by scaling the length of the center bulb towards the length and draught of the ship. The calculation can be seen in Table 3-4 and Figure 4.

Table 3. Main dimension of Danisman's catamaran

Lwl	2.525 m
B	0.26 m
T	0.14 m
Cb	0.41

Table 4. Calculation of center bulb model

Danisman Ellipsoidal Center Bulb	Scale	Ellipsoidal Center Bulb Experiment
Length (a) = 0.1414 m	0.1414/2.525=0.0558	Length (a) = 0.089 m~ 0.090 m
Width (b) = 0.062 m	0.062/0.26 =0.2384	Width (b) = 0.037 m
Height (c) = 0.034 m	0.034/0.14 =0.02428	Height (c) = 0.024 m
		Diameter (d) = (b+c)/2 = 0.0305 ~ 0.040 m

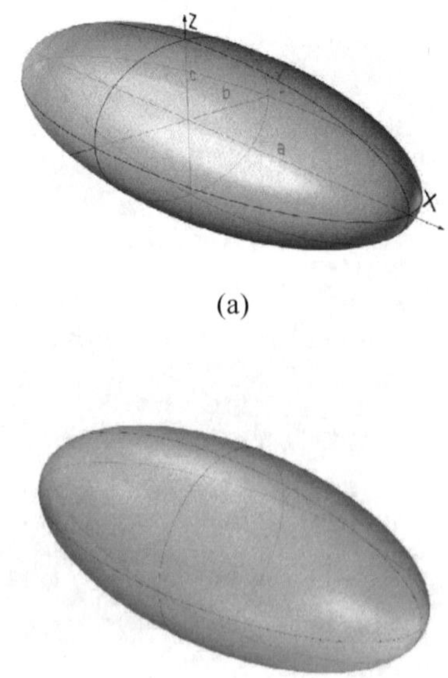

(a)

(b)

Figure 4. 3D geometry of Danisman center bulb(a) and center bulb experiment (b)

To place the center bulb, it positioned in the middle of the length of the waterline and the draught. The experiment is conducted on Fr 0.7 (2.7707 m/s). Based on the previous research, at this velocity, the catamaran with center bulb has a lower resistance compared to the catamaran without center bulb.

2.2. Experiment Using CFD

On how to find the resistance value, this research used Computational Fluid Dynamics (CFD) software. The software is called Tdyn and can be downloaded freely, which has its limitations if it is not registered. Similar to other CFD software, it has three main processes start from pre-processor, processor, and post-processor.

2.3. Design of Experiment (DoE)

In terms of simulating the ship's resistance in a CFD software, it is imperative to design the optimum center bulb by using Response Surface Methodology (RSM). Start from the first procedure, Order I, is to determine the minimum and maximum boundaries of the length (X1) and diameter (X2). The next step is Steepest Descent; this procedure is to find the value of X1 and X2 from the linear equation of Order I. The goal is to predict the lowest slope of the wave resistance. Results from Steepest Descent are used to make an initial model in Order II, which has the same purpose as Order I.

3. EXPERIMENT AND RESULTS

3.1. CFD Model Validation

Validation has a function to set similar conditions among the EFD model and CFD model. There are three variables to set, which are the number of steps, initial steps, and time increment. Those three variables are affecting the result of CFD software, which is total resistance (Rt).

There is a standard to validate the model, and this research using the margin of error below 1%. By using that standard, the model is considered uniform compared from EFD to CFD. However, before obtaining the value of the margin of error, the 3D model needs to have meshed.

Table 5. Mesh conditions

Mesh Condition	
Hull Surface	0.005
Free Surface	0.05
Other Surfaces	0.1
Max Element	0.5
Transitional	0.5

Mesh condition is obtained by trial and error based on the consideration to achieve the perfect 3D model. From that on, the main variables of the number of steps, time increment, and initial steps are obtained gradually. From this validation process, the result is Rt that consists of Rw (13.855 N) and Rv (6.0654 N)

Table 6. CFD conditions

Fr	0.7
Number of Steps	900
Time Increment	0.08
Initial Steps	81
V (m/s)	2.77
Rt (CFD)	19.11
Rt (Molland)	18.89
Error	0.12%

3.2. Response Surface Method (RSM)

3.2.(a) Order I

Based on the references, there are multiple kinds of Design of Experiments (DoE). The DoE used here is using the Center of Composite Design (CCD) with two variables of length (X1) dan diameter (X2) from the center bulb. Therefore, the linear equation result from Order I is the relation between X1 and X2. Variation difference applied to X1 and X2 is ±5% from the initial model.

Table 7. DoE based on CCD and Rw result.

Model	Code		Parameter		Rw
	X1	X2	L (mm)	D (mm)	
1	0	0	90.00	40.00	14.750
2	-1	-1	85.50	38.00	13.821
3	-1	1	85.50	42.00	14.510
4	1	-1	94.50	38.00	14.388
5	1	1	94.50	42.00	15.286

From Table 7, the linear equation is obtained by using the regression method. This equation has $R^2 = 0.94820$, where this value indicates the equation is valid.

$$RW = 0.3358X1 + 0.3969X2 + 14.5509 \quad (1)$$

3.2.(b) Steepest Descent

This procedure function is to find the lowest point of any value prediction of wave resistance (Rw) that tends to decline and no longer fit to calculation prediction from Equation 1. In this step, the addition or delta (Δ) of the coefficient of every factor from Order I calculated. Coefficient from X1 used as the initial value, therefore new equations obtained as shown in Equation 2 and Equation 3.

$$\Delta X1 = \frac{0.3358}{0.3358} = 1 \quad (2)$$

$$\Delta X2 = \frac{0.3969}{0.3358} = 1.1819 \quad (3)$$

Table 8. Steepest Descent

Model	Code		Parameter		Rw (Prediction)	Rw (CFD)
	X1	X2	L (mm)	D(mm)		
1	0.00	0.00	90.00	40.00	14.551	14.750
6	-1.00	1.18	85.50	42.36	14.684	15.029
7	-2.00	-2.36	81.00	35.27	12.941	13.111
8	-3.00	-3.55	76.50	32.91	12.136	14.031
9	-4.00	-4.73	72.00	30.54	11.331	14.040
10	-5.00	-5.91	67.50	28.18	10.526	14.041
11	-6.00	-7.09	63.00	25.82	9.721	13.499

Figure 5. Graph of Steepest Descent

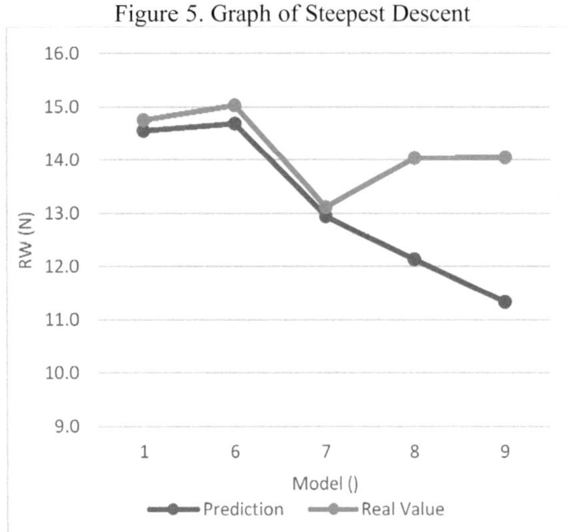

From the steepest descent experiment, the lowest point from the slope is on the Model 7 while the wave resistance is increasing on Model 8. Accordingly, Model 7 is used as initial geometry at Order II.

3.2.(c) Order II

In Order II, the process will generate a nonlinear equation, which in this term is a quadratic equation. It caused the addition of DoE, as shown in Table 10. Along with the initial value (X1=0, X2=0) is obtained based on the steepest descent process from Table 8 (Model 7). In that sense, it is imperative to redefine the code value in Order II. Table 9 shows the redefinition of its code.

Table 9. Code in Order II

	-1	0	1
X1 (Length, mm)	76.95	81.00	85.05
X2 (Diameter, mm)	33.51	35.27	37.04

Table 10. DoE based on CCD and Rw result

Model	X1	X2	X1*X1	X2*X2	X1*X2	L	D	Rw
7	0.00	0.00	0.00	0.00	0.00	81.00	35.27	13.11
12	-1.00	-1.00	1.00	1.00	1.00	76.95	33.51	14.29
13	-1.00	1.00	1.00	1.00	-1.00	76.95	37.04	13.14
14	1.00	-1.00	1.00	1.00	-1.00	85.05	33.51	14.19
15	1.00	1.00	1.00	1.00	1.00	85.05	37.04	14.37
16	-1.41	0.00	1.99	0.00	0.00	75.27	35.27	14.35
17	1.41	0.00	1.99	0.00	0.00	86.73	35.27	13.20
18	0.00	-1.41	0.00	1.99	0.00	81.00	32.78	13.27
19	0.00	1.41	0.00	1.99	0.00	81.00	37.77	14.50

Table 9 is used as the basis for DoE CCD, as shown in Table 10. As a result, the equation from Order II has R^2=0.3696. This value means that the equation not ideal to be used in finding the optimum geometry of the center bulb because the R^2 value did not come near to the value of 1.

$$RW = -0.061X1 + 0.096X2 + 0.373X1^2 + 0.430X2^2 + 0.329X1X2 + 13.1108 \quad (4)$$

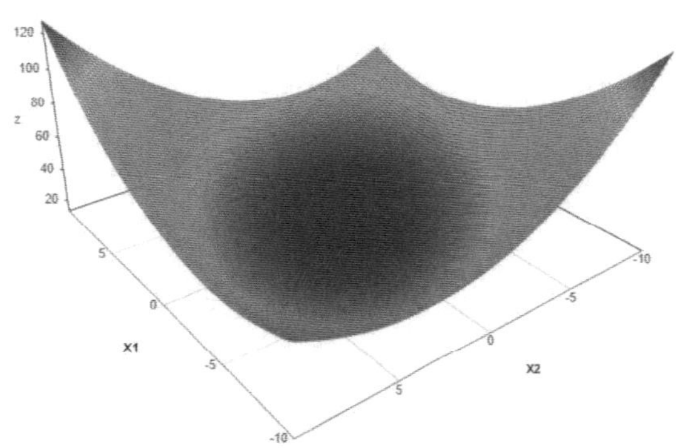

Figure 6. 3D graph of Order II

3.2.(d) Optimum Value

In order to find the minimum value from Order II (Equation 4), therefore it needs to find the first derivative from Equation 4. This process is shown in Equation 5 and Equation 6.

$$\frac{dRW}{dX1} = 0 \quad (5)$$

$$\frac{dRW}{dX2} = 0 \quad (6)$$

From that process, the value of X1=0.517582608 and X2=-0.172143623. Those values are used to redefine the code in Table 9. However, when those values used in the Order II equation, it predicted the Rw=13.0659 N and the real value from the experiment Rw=14.15 N (Model 20). The Model 20 dimension is shown in Table 11 and Figure 7, with the comparison with Model 1.

Table 11. Main dimension of Model 20

Model	Code		Parameter	
	X1	X2	L (mm)	D(mm)
20	0.158	-0.172	81.63	34.96

Figure 7. Comparison of Model 1 (continuous line) and Model 20 (dashed line)

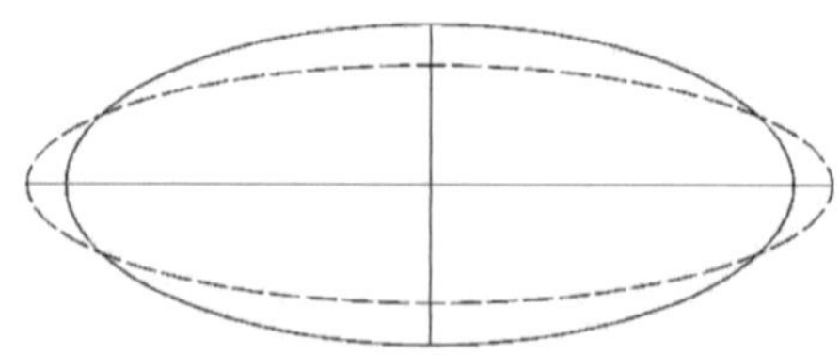

The resulting design generates wave resistance bigger than the catamaran without a center bulb installed. It concludes that the Model 20 doesn't suit the criteria to reduce wave resistance compared to Model 7 (Rw=13.111 N). In that sense, Model 7 is considered the optimum model, and it can minimize wave resistance by 5.36% compared to catamaran without the center bulb.

Table 12. Main dimension of Model 7

Model	Code		Parameter	
	X1	X2	L (mm)	D(mm)
7	-2	-2.364	81	35.27

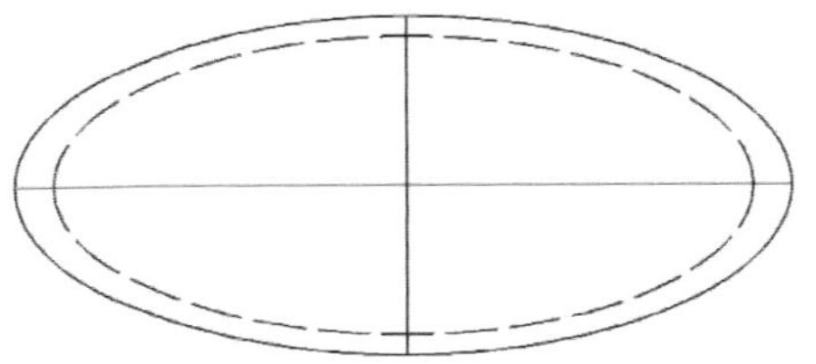

Figure 8. Comparison of Model 1 (continuous line) and Model 7 (dashed line)

Table 11 shows the dimensions of Model 7, and Figure 7 is a comparison between Model 1 and Model 7. From that, it can understand that Model 7 has a smaller geometry compared to Model 1, in which the length from 90 mm changes to 81 mm or reduce by 10%, and the diameter change from 40 mm to 35.27 or reduce by 11.82%.

3.3. Hydrodynamic Comparison

A center bulb without optimization, Model 1, has a wave resistance (Rw) of 14.750 N. This value is bigger by 6.45% compared to the catamaran model without a center bulb. It caused by the size of the geometry generates bigger interferences. This phenomenon is shown in Figure 9b, in which it shows that the pressure concentration is located at the tip of the center bulb and adding more pressure to overall wave resistance.

In comparison, Model 7 (optimized center bulb) can reduce the wave resistance by 5.36% compared to the model without a center bulb. As shown in Figure 9d, the pressure distributed evenly in between the hulls. It concludes that the Model 7 achieved the goal to reduce the wave resistance.

However, Model 20 (optimization from Order II) can't achieve the goal to reduce the wave resistance. Even though the prediction generated Rw=13.0659, but the real wave resistance value is 14.15 N. It concludes that Model 20 can't reduce the wave resistance (Figure 9c).

Compared to Danisman's research, Rw reduce by 5.36% considered not significant. This result is probably caused; there are only two variables being researched (length and diameter) while Danisman was also considering the location of the center bulb. It played a significant role in reducing the wave resistance.

Overall, this research proved that the Response Surface Methodology (RSM) could help to optimize the geometry of the center bulb. It also means that RSM could be used in broad topics of Naval Engineering. Besides that, it gives an easier approach for researchers because it doesn't need a wide variety of DoEs.

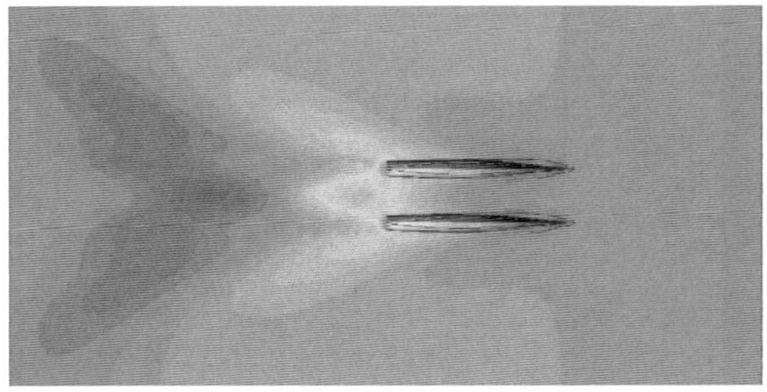

(a)

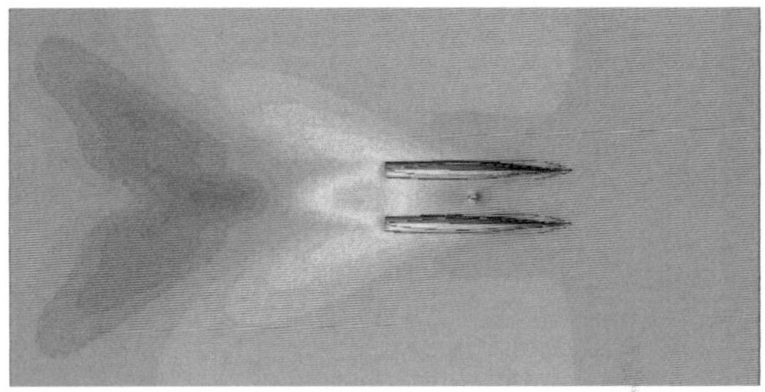

(b)

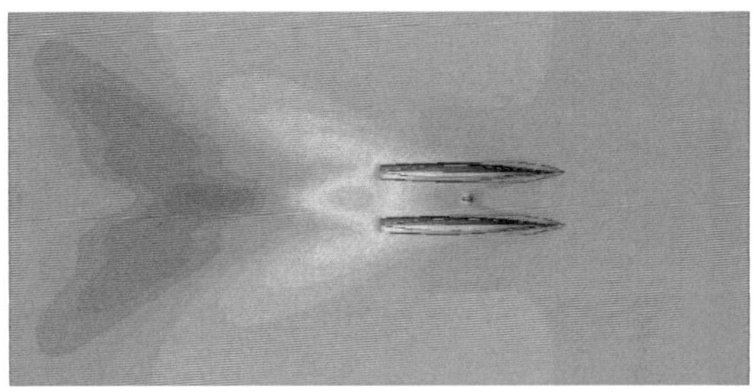

(c)

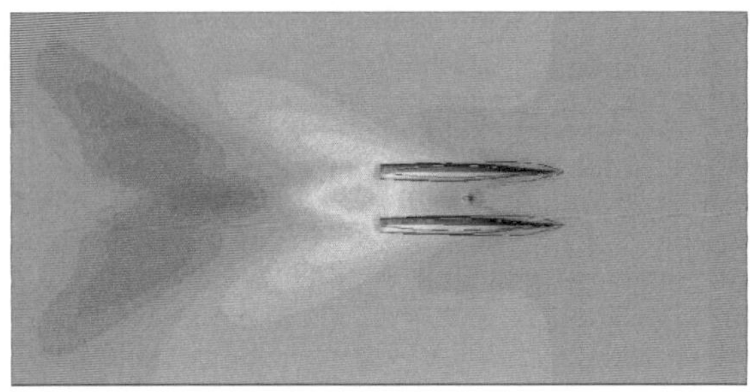

(d)

Figure 9. Comparison of catamaran without center bulb (a), Model 1 without optimization (b), Model 20 with Order II optimization (c), and Model 7 with Steepest Descent optimization(d)

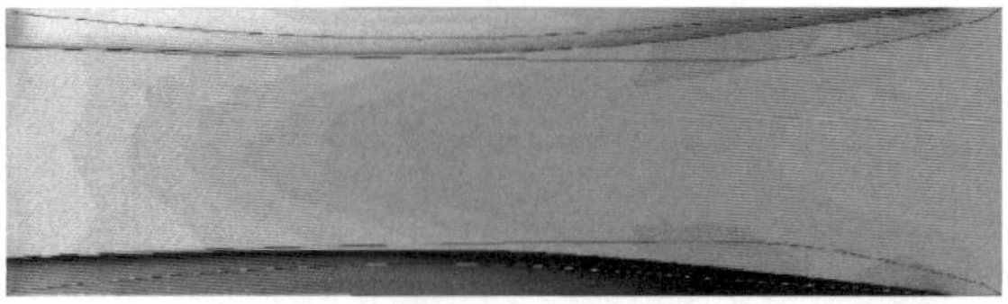

(a)

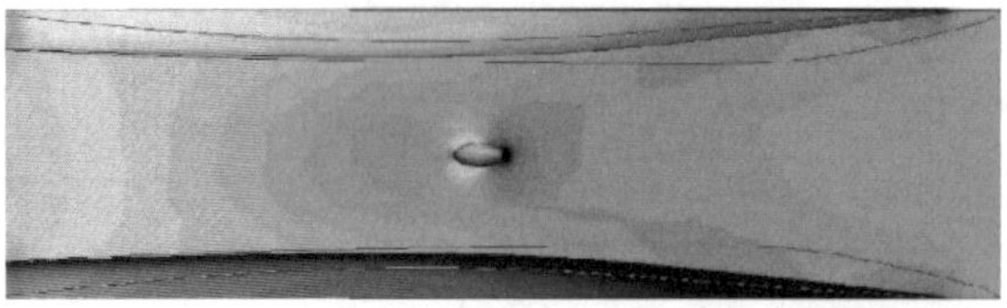

(b)

(c)

Figure 10. Detailed comparison of catamaran without center bulb (a), Model 1 without optimization (b), Model 20 with Order II optimization (c), and Model 7 with Steepest Descent optimization(d)

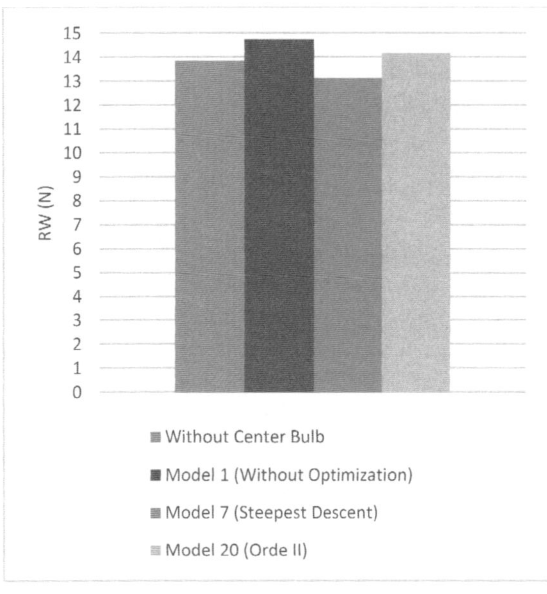

Figure 11. Rw comparison for each center bulbs

4. CONCLUSIONS

The optimum geometry is 81 mm in length from 90 mm or decreased by 10% and 35.27 mm in diameter from 40 mm or decreased by 11.82%. This model can reduce wave resistance (Rw) by 11.11% compared to the catamaran with a non-optimized center bulb. It also can reduce by 5.36% of wave resistance compared to catamaran without center bulb. Overall, this research proved that RMS could be used in Nava Engineering topics.

5. ACKNOWLEDGMENTS

The writers would like to thank the Faculty of Engineering of Diponegoro University for their support towards this research.

6. REFERENCES

1. E. Jahanbakhsh, R. Panahi and M. S. Seif, "CATAMARAN MOTION SIMULATION BASED ON MOVING GRID TECHNIQUE," *Journal of Marine Science and Technology,* vol. 17, no. 2, pp. 128 - 136, 2009.

2. F. Zouridakis, A Preliminary Design Tool for Resistance and Powering Prediction of Catamaran Vessels, Boston: MIT, 2005.

3. Samuel, M. Iqbal and I. Utama, "An investigation into the resistance components of coverting a traditional monohull fishing vessel into catamaran form," *International Journal of Technology,* vol. 6, no. 3, pp. 432-441, 2015.

4. Samuel, D.-J. Kim, M. Iqbal, A. Bahatmaka and A. R. Prabowo, "Modification of traditional catamaran to reduce total resistance: Configuration of centerbulb," in *International Maritime Association of the Mediteranean (IMAM)*, Lisbon - Portugal, 2017.

5. M. Iqbal and G. Rindo, "Optimasi Bentuk Demihull Kapal Katamaran Untuk Meningkatkan Kualitas Seakeeping," *KAPAL,* vol. 12, no. 1, pp. 19-24, 2015.

6. S. Samuel, D.-J. Kim, M. Iqbal, A. Bahatmaka and A. R. Prabowo, "Bulbous bow applications on a catamaran fishing vessel for improving performance," *MATEC Web of Conferences,* vol. 159, 2018.

7. D. B. Danisman, "Reduction of demi-hull wave interference resistance in fast displacement catamarans utilizing an optimized centerbulb concept.," *Ocean Engineering,* vol. 91, pp. 227-234, 2014.

8. I. Zotti, "Medium speed catamaran with large central bulbs: experimental investigation on resistance and

vertical motions," in *ICMRT'07*, Ischia, Naples, Italy, 2007.

9. R. P. Aprjal, S. Samuel and M. Iqbal, "Minimisasi Hambatan Dan Gerak Vertikal Kapal Multihull Catamaran Dengan Centerbulbs," *Teknik,* vol. 39, no. 1, 2018.

10. D. Bruzzone, A. Grasso and I. Zotti, "Nonlinear Seakeeping Analysis of Catamarans with Central Bulb," in *6th International Conference on High-Performance Marine Vehicles.*, 2008, 2008.

11. G. K. Saha, K. Suzuki and H. Kai, "Hydrodynamic Optimization of a Catamaran Hull with Large Bow and Stern Bulbs Installed on the Center Plane of the Catamaran," *Journal of Marine Science and Technology,* vol. 10, pp. 32-40, 2005.

12. M. A. Bezerra, R. E. Santelli, E. P. Oliveira, L. S. Villar and L. A. Escaleira, "Response surface methodology (RSM) as a tool for optimization in analytical chemistry," *Talanta,* vol. 76, pp. 965-977, 2008.

13. L. Ma, Y. Han , K. Sun, J. Lu and J. Ding, "Optimization of acidified oil esterification catalyzed by sulfonated cation exchange resin using response surface methodology.," *Energy Convers Manage,* vol. 98, p. 46–53, 2015.

14. A. Baroutaji, M. D. Gilchrist , D. Smyth and A. Olabi, "Crush analysis and multi-objective optimization design for circular tube under quasi-static lateral loading," *Thin-Walled Structures,* vol. 86, pp. 121-131, 2015.

15. O. I. Awad, R. Mamat, O. M. Ali, W. H. Azmi, K. Kadirgama, I. M. Yusri, A. M. Leman and T. Yusaf, "Response surface methodology (RSM) based multi-objective optimization of fusel oil -gasoline blends at different water content in SI engine," *Energy Conversion and Management,* vol. 150, pp. 222-241, 2017.

16. A. F. Molland, S. R. Turnock and D. A. Hudson, Ship resistance and propulsion, Cambridge university press, 2017.

7. AUTHORS BIOGRAPHY

M. Iqbal. Lecturer at Department of Naval Architecture, Diponegoro University. His previous research is about Optimization of Foil-Shaped Centre Bulb using Response Surface Method.

E. S. Hadi. Lecturer at Department of Naval Architecture, Diponegoro University. Head of Hydrodynamics Laboratory Diponegoro University. He is also Doctoral Student in Departement of Mechanical Enginering Diponegoro University

G. Pranamya. Undergraduate Student of Department of Naval Architecture, Diponegoro University. He also researcher member of previous research about Optimization of Foil-Shaped Centre Bulb using Response Surface Method

TOWARD GREEN AND SUSTAIANABLE SHIP RECYCLING INDUSTRY IN INDONESIA

Sunaryo, Naval Architecture and Marine Engineering Study Program Universitas Indonesia, Indonesia
E B Djatmiko, Department of Marine Engineering Institut Teknologi Sepuluh Nopember, Indonesia
R E Kurt, Department of Naval Architecture Ocean and Marine Engineering University of Strathclyde, UK

SUMMARY

The main aim of this project is to develop a sustainable, safe & environmentally sound ship recycling industry in Indonesia. As the world's largest archipelago state, it is essential for Indonesia to maintain a large fleet of ships to ensure sufficient sea transportation. However, after completing their operational life, all these ships are required to be dismantled. Currently, there is no proper understanding & infrastructure for 'Ship Recycling' in Indonesia yet. This lack of understanding causes serious contamination of the sea which results in harming the environment. Furthermore, the working conditions are extremely poor because the yards are not fit for purpose and workers are exposed to many hazards. Referring to International Maritime Organization's Hong Kong Convention a project is carried out to preparing the process of establishing green and sustainable ship recycling industry in Indonesia by conducting gap analyses and proposing environmentally friendly ship recycling yard concept.

NOMENCLATURE

DO	Dissolved oxygen
EU	European Union
GT	Gross Tonnage
IDR	Indonesian Rupiah
IISIA	Indonesian Iron and Steel Industry Association
IMO	International Maritime Organization
Pb	Lead
PCB	polychlorinated biphenyl
pH	Concentration of hydrogen ions in the solution
PPE	personal protective equipment
ppm	parts per million
SNI	Standar Nasional Indonesia
UNEP	United Nations Environment Programme
USD	United States Dollar

1. INTRODUCTION

The project is conducted as part of collaborative study between Universitas Indonesia, Institut Teknologi Sepuluh Nopember, and University of Strathclyde under the funding of the Newton Fund Institutional Link and Ministry of Research Technology and Higher Education Republic of Indonesia Basic Research Funding. With the objective to develop a sustainable, safe & environmentally sound ship recycling industry in Indonesia.

Indonesia is the world's largest archipelago state with more than 17,000 islands. Therefore, it is important for Indonesia to maintain a large fleet of ships to ensure sufficient sea transportation. Hence, the government gave national shipping companies incentives to import used ships from abroad, which resulted in great number of older ships operating in Indonesia. However, these old ships when reach their uneconomical stage are required to be dismantled and replaced by new fleet. The number of Indonesian flagged shipping fleet increased significantly since the implementation of Cabotage principle through the Presidential Instruction No.5/2005 [1], from 6,041 units in March 2005 to 24,026 units in June 2016 [2]. According to the statement from Indonesian National Ship Owners Association the increase is mainly due to import of used ships from abroad, and around 70% of the fleet are categorized as old ships, which is ranging from 15 to 30 years of age [3], and according to Roesdianto [4] 20% of them have reached more than 25 years. Refer to ship life cycle norms, these uneconomical for operation ships need to be replaces with the new ones, and sent to the ship recycling yards.

The term 'ship recycling' has just been adopted since the establishment of International Maritime Organization's the Hong Kong International Convention for the Safe and Environmentally Sound Recycling of Ships (2009) [5]. Ship recycling is defined as an activity that dismantle ships partly or entirety in a facility which goal is to retrieve the components and material from the ships to be processed or reused while handling the waste in different waste treatment facility (IMO, 2009), ship recycling is reckoned not similar to ship scrapping activity; ship scrapping or ship breaking only focusing on dismantling the ship without considering their effects to environment, health and safety of the community (Gwin, 2014) [6].

With regard to the great number of old ships exist in Indonesia, it is assumed that the sustainability of ship recycling market is assured; moreover some of the developed leading shipping countries such as Japan, Korea, China, and European countries prefer to send their uneconomical ships for recycling abroad such as to Bangladesh, Pakistan, or India [7]. This trend would also be a potential market for Indonesia's ship recycling industry.

Currently there is no proper understanding and infrastructure for 'Ship Recycling' in Indonesia yet. Some rather environmentally aware ship recycling activities are carried out by ship repair yards in Batam and Jawa Islands, but only for a few number of ships. Most of the ship recycling activities is conducted traditionally by beaching the ships to the shore and using simple tools and equipments. Parts and components that have been stripped off from the ship are temporally stacked on the shore waiting to be transported for further processing or directly sold to the end buyers. All kind of pollutants are usually found on the beach and the surrounding water fronts, which are actually threat to the environment and health of the neighbouring communities. These traditional ship recycling yards are concentrated in Bangkalan, south-western part of Madura Island, and in Cilincing North Jakarta. The existence of them are categorized as 'Semi-legal', because they are not legally registered to the related authorities, even though their activities are relatively very easy to be spotted, therefore precise data related to the activities are very difficult to obtain.

Since ship recycling is always associated with hazardous and toxic wastes, therefore strict regulations have been established both internationally as well as nationally. Some of these regulations are: international regulations: Hong Kong International Convention for the Safe and Environmentally Sound Recycling of Ships (IMO 2009); The Basel Convention on the Control of Tran boundary movements of Hazardous Wastes and their Disposal (UNEP 1989) [8]; EU Waste Shipment Regulation (Regulation (EC) No 1013/2006) [9]; EU Ship Recycling Regulation (Regulation (EU) No 1257/2013) [10]. National regulations: Regulation of the Ministry of Transport Republic of Indonesia No. PM 29/2014 on the Prevention of Maritime environmental pollution. Chapter 51 – 56[11]; Shipping Law no. 17, 2008 [12], chapter 241, and 242 on ship recycling; Government regulation No. 101 year 2014 on treatment of dangerous and poisonous wastes [13].

Considering the potential sustainable ship recycling market for Indonesia, and backed up by the availability of international and national regulations, there are great opportunities to implement green and sustainable ship recycling industry in Indonesia, and based on this challenge the project is carried out.

2. METHODOLOGY

The following methodology is implemented in the project, started by indentifying the international and national requirements for sustainable and environmentally friendly ship recycling activities, and then gap analysis is conducted by comparing the requirements with the existing conditions of the ship recycling activities carried out in Indonesia. Based on the findings of gap analysis, concept of green ship recycling facilities are proposed, followed by developing supporting instruments for the implementation as illustrated in figure 1..

2.1 REQUIREMENTS FOR GREEN SHIP RECYCLING PROCESS

Hong Kong International Convention for the Safe and Environmentally Sound Recycling of Ships requires that ship recycling process should comply with its regulations as follows: The convention is applied to commercial ships above 500GT, Ship owner shall finalize the Inventory of Hazardous List and based on the information from ship owner, ship recycler shall provide Ship Recycling Plan, and get approval from the authority, Final Inspection by flag nation, Issue of International Certificate of Ready for Recycling, conduct the recycling, notation of completion of recycling to the flag nation and recycling nation.

Basel Convention on the technical guidelines on Environmentally Sound control management and dismantling of ship parts, which provide information and recommendations on procedures, processes and practices that must be implemented to attain Environmentally Sound Management at such facilities

On the national level Regulation of the Ministry of Transport Republic of Indonesia No. PM 29/2014 on the Prevention of Maritime environmental pollution, the regulation on commercial ships above 500GT is exactly the same as Hong Kong Convention, but it also cater for ships of 100GT and above, including the procedures for recycling processes and ship recycling facilities.

Shipping Law no. 17, 2008 chapter 241, and 242 regulate the location for ship recycling, and protection of environment for ship recycling.

Government regulation No. 101 year 2014 on treatment of dangerous and poisonous wastes regulate the procedures of waste treatment from collection, transporting, treatment, stacking, dumping, and inspection.

2.2 REVIEW ON SHIP RECYCLING METHODS

2.2 (a) Beaching

It is the most method applied in many developing countries with significant tidal range, and sloping beach. When the tide is high, the vessel is driven towards the beach or even grounded, once the water recedes first the workers will empty the ship, then the ship is progressively broken in to smaller sections manually using simple tools such as oxy-acetylene flame cutters. The ship parts and accessories are brought to the dry land. This method is cheap because it does not need any crane or heavy winch to bring the ship ashore, nor dry dock or slipway. The pollutants and hazardous materials that are spilled out from the wastes are usually spreader on shore, absorbed by the ground, or flow to the sea, hence it is the most unenvironmentally friendly method.

2.2 (b) Landed

Almost similar to beaching, landed is usually applied where the tidal range is not very high. The ship is grounded to the beach during the high tide, and then when the tide is low front, higher and heavy parts of the ship are cut off, while the aft part is still floating, when the water is up again the ship is pulled farther closer to the beach. This operation is continuously carried out until the whole ship is totally dismantled. The environmental risks of this method are almost the same as beaching, but the safety risks are even higher because of the movement of the ship during the dismantling process.

2.2 (c) Alongside

It is known also as quayside, pier side or floating method, the ship is dismantled afloat and moored along wharfs, jetties or quays. Heavy cutting gear is used to reduce the ship from the top parts first, and the ship is cleared from its loose items. Then the last empty floating hull is reduced to the extent possible while afloat and then either taken out as a whole or further cut into pieces in a dry dock. The environmental risks are less compared to beaching, but the safety risks are as high as landed and more expensive.

2.2 (d) Dry-docked

The ship is towed to a dry-dock, or floating dock which has a lock gate and an impermeable floor structure. The dock is usually provided with crane and other equipments that comply to ship recycling facility plan. The ship is placed on the docking stools in accordance with a docking plan. Pollution to the environment can be reduced to almost nothing because it allows for full containment. Heavy and automated tools and equipment are used. It is the most environmentally sound and safe way of ship recycling, but also the most costly method.

2.2 (e) Slipway

This can be considered as the compromised way of landed and dry-docked. The ship with all of its liquid contained had been emptied is towed to the recycling yard, then placed on the rail mounted buggies, the buggies are pulled by a winch to the dry concrete covered floor. The dismantling process is the same as that of dry-docking method. The environmental and safety risks of this method are not as low as dry docking, because the pollutants can still spill out to the sea, and the ship is not as stabile as on the dry-dock.

2.3 SHIP RECYCLING IN INDONESIA

Most of the ship recycling activities carried out in Indonesia are using beaching method, only a handful number of ship repair yards implementing more Hong Kong convention compliant method ship recycling, and the number of ships have been recycled also very few, slipway recycling method is usually implemented by these yards.
The most popular place for beaching ship recycling is in Bangkalan, south-western region of Madura Island in East Jawa, and the second one is in Cilincing North Jakarta. The ship recycling yards in these two places are reckoned as 'Semi-legal', because even though the activities are very easily identified, but the yards are not officially registered to, and poses legal certificates from the related authorities, such as local government, Ministry of Environment and Forestry, Ministry of Industry etc.
Negative effects of these 'Semi-legal' ship recycling activities are very high for environment, health and safety of people directly involved in the activities or those who live in the neighbouring area. The hazardous wastes are not treated properly, there is not even waste treatment facility exists in the regions. Wastes such as dirty liquid are directly discharged to the sea, and oil is collected in the containers to be sold and reused in other industries, paint/anti fouling, rust are just scattered on the ground. Other hazardous solid wastes such as asbestos, PCB, glasswool, heavy metals etc. People who work in the yards are exposed to hazardous materials and dangerous environment, not all of them wear proper personal protective equipment (PPE), and very limited safety equipments and emergency measures.

2.4 GAP ANALYSIS

In order to define the strategy for proposing green and sustainable ship recycling industry in Indonesia, a gap analysis is carried out by comparing both international as well as national requirements and the existing condition of the ship recycling industry in Indonesia.
Business and environmental surveys were conducted on 'Legal' and 'Semi-legal' ship recycling activities in Batam, Banten, Cilincing, and Madura. Objects of survey for business activities are: demand and market, the activity flow, employment system and workers' welfare. For the environmental impacts are: types and level of contamination on the shore, in the water, and in fishes.

2.4 (a) Demand and market analysis

Demand from steel mill industry for iron steel scrap is very high in response to the Government policy in boosting the infrastructure projects. Based on the information given by The Indonesian Iron and Steel Industry Association (IISIA) [14] that raw material from steel scrap for iron and steel industry are 70% imported from abroad, and 30% are obtained locally, the total need of scrap steel are approximately around 6.5 million tonnes per annum. In the last two years the price of this raw material has increase from USD.450 to USD.800 per ton, due to lack of supply. This high demand of steel can be fulfilled from ship recycling industry.
The demand for used spare parts and machinery from the existing great number of used ships for replacement purposes is also very high, because certain prepares might not be available in the market anymore, therefore spare parts and machinery from recycled ships are the only source for fulfilling the demand.

As mentioned in the introduction, since the implementation of cabotage principle and importing used ships incentive by Indonesian Government, the number of used ships increased significantly. These ships gradually need to be replacing by the new ones, and thus the old and uneconomical ships are subject to recycling. Moreover many of developed countries are restricting ship recycling industry in their countries, and prefer to send their old ships to ship recycling facilities in developing countries, this also is becoming potential market for Indonesia.

2.4 (b) Activity flow

There are not definite information regarding the business chain of ship recycling but through information obtained from informal sources, the yards purchase ships to be recycled from ship brokers, who obtained the ships through various ways.
Usually the products of recycled ship are not directly sold by the yards to the end users, but through specialized brokers or intermediary buyers, who are always present around the ship recycling yards.
Loose items, equipments, and machinery of the dismantled ship are sold for replacement of parts and equipments of existing fleet. The straight and flat steel plates are cleaned and sold to ship repair yards. Other steel parts from ship structure are sent to steel mills as raw material for other steel products.

2.4 (c) Employment system and welfare analysis

In the 'modern' yards, which are actually ship repair yards permanent or contract workers are employed. The employment system of these workers follows the existing industrial relation regulations, including their waging system, health insurance, and other welfare system.
In the traditional (Semi-legal) ship recycling yards the workers are not permanently employed, who come from the neighbouring areas. They work as daily paid workers; the number of them is fluctuating depended on the work availability, and the type and size of ship being recycled. For example yard one 1,000 Dwt general cargo ship there are around 20 to 30 workers employed.
The wages of workers are ranging from IDR 250,000 for those dismantling block structure, pipes and machinery, and heavy equipment operators, to IDR 90,000 for plate cleaning and cable stripping workers per day. These daily paid workers are not insured, nor receive any other income beside their daily salary.

2.4 (d) Environmental analyses

In the 'Legal' yards the wastes are collected in accordance to their categories in separate containers. To dispose the wastes subcontractors are usually employed, and there are no clear information where and how the wastes are treated.
There is no proper waste handling and wastes treatment facilities found during the survey in the 'Semi-legal' ship recycling yards both in Cilincing and in Madura.
Since most of the ship recycling activities is carried out in western part of Madura Island more precisely in Kabupaten Bangkalan, the environmental samples were taken in three neighbouring places i.e. Bangkalan, Kamal, and Socah. The samples taken are water (taken from bottom and from surface), and sediment from the same location at the depth of 20cm, in five randomly selected points from each location, the distance from one point to another is about 15 m. Three locations were surveyed
The target of investigation of the samples are: from the water: salinity, pH, temperature, and dissolved oxygen; and from the sediment is Pb.
Results of the samples been collected are:

Location	Pb content (ppm) Water average	Sediment average
1	0.24	0.703
2	0.07	0.747
3	0.09	0.788

Location	Water quality Temperature (^{0}C)	pH	DO	Salinity
1	30	7	5	35
2	28	7	5	33
3	29	7	6	31

Based on the decree of Minister of Environment No. 51 (2004) [15] the accepted standards Pb content is 0.008 ppm, temperature 28 to 32 ^{0}C, pH 7 to 8.5, DO >5, salinity 33 to 34. Thus the contamination of Pb to both water and sediment are above the permitted level, but the water quality is still in the permitted levels.
Data collected for the food chain contaminated analysis are: The average content of heavy metals of lead (Pb) on puffer fish are 0.043 ppm, on manyung fish 0.095 ppm, on white shrimp 0.103 ppm, on crab 0.113 ppm, squid 0.149 ppm, and mussels 0.08 ppm. The content has not exceeded the threshold value set by the government (fish: 0.3 ppm; crustaceans: 0.5 ppm and mollusks: 1.5 ppm) (SNI, 2009) [16] so it is still safe for human consumption.

2.5 PROPOSED DEVELOPMENT OF GREEN SHIP RECYCLING

Based on the results of the gap analysis, strategy for developing green and sustainable ship recycling industry in Indonesia is proposed. Three strategic options are proposed i.e.: 1. Transforming the existing 'Semi-legal' ship recycling yards into green ship recycling yards; 2. Introducing airbag green ship recycling concepts to any potential new businesses; 3. Establishing integrated green ship recycling industry cluster.

2.5 (a) Transformation of existing ship recycling yards

The 'Semi-legal' ship recycling yards in Madura and Cilincing have been existed for more than two decades, and provided jobs and living for hundreds of households, even though their existence is harming to the environment and health of the people surrounding the areas. It is very difficult to force them to stop the business or change to environmentally friendly methods without educating and equipping them with proper approaches. One of the approaches is through gradual transformation of the ship recycling practices unto green recycling yards. Initiative should come from the authority such as local Government by providing assistance to the ship recycling owners to convert the facilities to be more compliant to the regulations with favourable financing scheme, and legal certifications; developing supporting infrastructure such as transportation roads, waste treatment facilities, community health centres, simple housing for the workers, training centre etc., and periodically having discussion forums between the stakeholders for transforming their mindset to adopt the green ship recycling concepts.

2.5 (b) Introduction of air bag green ship recycling concepts

Since the ship recycling market is so promising for Indonesia, beside the existing 'Semi-legal' ship recycling yards that should be transformed into green ship recycling yards, there are also high potential for developing new, or converting ship repair yards into 'modern and green' ship recycling yards as long as the regulations and business atmosphere are supporting.
In small and middle sizes shipyards airbags are extensively being used. Therefore instead of implementing dry-docking, slipway, and alongside, airbags rolled on the concrete slop ground method is introduced. In order to comply with the regulations, all the requirements should be fulfilled, starting from preparation until approval of procedures reporting, the only difference of this method is only the docking process of the ship. After arriving at the yard during the high tide, it is pulled by means of winch to the slope inclining ground. Prior being pulled the ship is aft trimmed by ballasting, then airbags are inserted under the front bottom of the ship, this procedure is carried on so as the ship rolls on the airbags, until the whole ship sits pulled for a certain distance from the water on the dry ground. The airbags are then replaced by docking stools for supporting the ship, and then end of the inclining ground is blocked by using concrete portable separators to prevent wastes and contaminants spill over to the sea. The rest of the following procedures are the same as implemented in dry-docking method. Illustration of this method is presented in figure 2.

2.5 (c) Integrated green ship recycling industry cluster.

In order to achieve high efficiency in logistic system of the ship recycling industry, and they have greater control of the treatment of hazardous materials Integrated green ship recycling industry cluster is proposed.
All kind of business chain related to the ship recycling industry is suggested to be located in the cluster, and each of the business is connected through an agglomeration networks. The business that should be included in the cluster are: ship recycling yards; used-ship brokerage; waste treatment facilities; steel mill; hospital or health centre; worker housing; worker training centre; ship used components market, flame cutting fuel distributors; fresh water and electricity suppliers; authority institutional offices etc. Some considerations have been put forward in proposing the location for the integrated green ship recycling cluster, including: the existing largest ship recycling place, such as Madura Island; un-productive shore base or island but has suitable coast line; close to the busy shipping lanes, so that old ships can be easily driven to the yards.

3. RESULTS AND DISCUSSION

Since the implementation of Cabotage principle in 2005 the number of national fleet in Indonesia has been increasing significantly due to the incentives from Government to national shipping firms to purchase used-ships from abroad. But 70% of the fleet are consisted of old ships, even 20% of which are now have reached more than 25 years, which soon should be recycled and replace the younger ones. Refer to these facts there are high demand for ship recycling. On the other hand demand for the scrapped steel is also high from the steel mills industry due to the boosting of infrastructure projects in Indonesia.
Majority of ship recycling activities carried out in Indonesia is through beaching the ships to the shore. The activities are mainly concentrated in south-western part of Madura Island namely in Bangkalan, and in Cilincing North Jakarta. There is no official number of these recycling yards, because they are reckoned as 'Semi-legal' businesses. More proper ship recycling yards are also existed in Batam Island, and in Banten, but their number are less than five, and the number of ships have been recycled are also very few.
Impacts of the activities of 'Semi-legal' ship recycling yards are very risky to the environment, and health and safety of the people who work and live in the neighbourhood. Based on the survey results the water and sediment in Bangkalan region of Madura has highly contaminated by heavy metal of Pb or lead, but the water quality is still considerably acceptable. Samples related to food chain taken from fishes are also in category of acceptable.
Since demand for ship recycling is high for Indonesia, and the related regulations are also available, it is concluded that sustainable and environmentally friendly ship recycling industry in Indonesia has great potential opportunity, even though some applicable approaches need to be considered.

4. COCLUSIONS

Implementation of green and sustainable ship recycling industry in Indonesia is feasible, because there is potential market of it both nationally as well as internationally. Related regulations are also available to guide the activities. But since most of the activities are still considered as 'Semi-legal', and their practices are not environmentally friendly, therefore many efforts are needed to achieve the goals of the project. In relation to the objective of the project three options to develop green and sustainable ship recycling industry in Indonesia have been proposed, include 1. Transforming the existing 'Semi-legal' ship recycling yards into green ship recycling yards; 2. Introducing airbag green ship recycling concepts to any potential new businesses; 3. Establishing integrated green ship recycling industry cluster.

5. ACKNOWLEDGEMENTS

The authors would like to express their gratefulness to the Newton Fund Institutional Links and the Ministry of Research Technology and Higher Education for providing funding to carry out the project both in Indonesia as well as in the UK.

6. REFERENCES

1 Secretary Of State Republic Of Indonesia, 'Presidential Instruction No.5 – 2005 on the Empowerment of National Shipping Industry' (*Indonesian*), *Secretariat of State Republic Of Indonesia,* 2005.

2 Saiful, 'The Number of National Ships is 24,026 Unit Per 2016' (Indonesian), *Indonesia Sipping Line,* http://www.indonesiashippingline.com/, 2018.

3 Hasbullah M., 'Development Strategy for National Shipbuilding Yards for Increasing Affectivity and Efficiency of National Shipping Fleet' (Indonesian), *Jurnal Riset dan Teknologi Kelautan vol. 14, no. 1*, 2016.

4 Roesdianto, T. 'Shipyard Management' (Indonesian). *IPERINDO*, 2017.

5 International Maritime Organization, 'Hong Kong International Convention for the Safe and Environmentally Sound Recycling of Ship', *IMO*, 2009.

6 Gwin, P., 'The Ship-Breakers', *National Geographic,* 2014.

7 Gasc E, 'The European List of Ship Recycling Facilities', *Directorate General Environment, European Commission*, 2018.

8 United Nation Environment Programme, 'Basel Convention on the Control of Tran boundary Movements of Hazardous Wastes and their Disposal', *UNEP*, 1989.

9 The European Parliament and the Council of the European Union, 'EU Waste Shipment Regulation (Regulation (EC) No 1013/2006)', Official *Journal of the European Union*, 2006.

10 The European Parliament And The Council Of The European Union, 'EU Ship Recycling Regulation (Regulation (EU) No 1257/2013)', *Official Journal of the European Union*, 2013.

11 Ministry of Transport Republic of Indonesia, 'Minister Regulation No. PM 29/2014 on the Prevention of Maritime environmental pollution' (Indonesian), Directorate of Sea Transportation, 2014.

12 Minister of Law Republic Of Indonesia, 'Indonesia Shipping Law no. 17, 2008' (Indonesian), Secretary Of State Republic Of Indonesia, 2008.

13 Secretary Of State Republic Of Indonesia, 'Government regulation No. 101, 2014 on treatment of dangerous and poisonous wastes' (Indonesian), Secretary Of State Republic Of Indonesia, 2014.

14 Pinem E, 'High Price of Raw Materials as Obstacle for Steel Industry' (Indonesia), Ministry of Industry, http://www.kemenperin.go.id/artikel/4812&t, 2019.

15 Minister of Environment, 'Decree of the Minister of Environment No. 51, 2004' (Indonesian), *Ministry of Environment*, 2004.

16 Jailani A Q and Kristiani M, 'Heavy Metals Pb In Water, Sediment And Shrimp Ghost Shrimp In The Coastal Subdistrict Labang, Bangkalan Madura', *Journal Of Aquaculture Development And Environment Vol.1, No. 1,* 2018.

7. AUTHORS BIOGRAPHY

Prof. Sunaryo, holds the current position of Director of Indonesian Maritime Center at Universitas Indonesia. He is responsible for coordinating researches, studies, and collaborations in any maritime fields. His previous experience includes ferry safety, environmentally friendly ship recycling process, environmentally friendly materials for boat building etc.

Prof. Eko Budi Djatmiko holds the current position of lecturer and superintendent of the Offshore Hydrodynamics Laboratory at Institut Teknologi Sepuluh Nopember. He is responsible for coordinating researches in the fields of ocean engineering and marine environment. His previous experience includes developing sustainable and green ship recycling facilities

Dr. Rafet Emek Kurt holds the current position of senior lecturer and principal researcher at University of Strathclyde. He is responsible for supervising PhD students, and conducting researches. His previous experience includes investigation of human factors in the maritime domain, and research in ship recycling sectors.

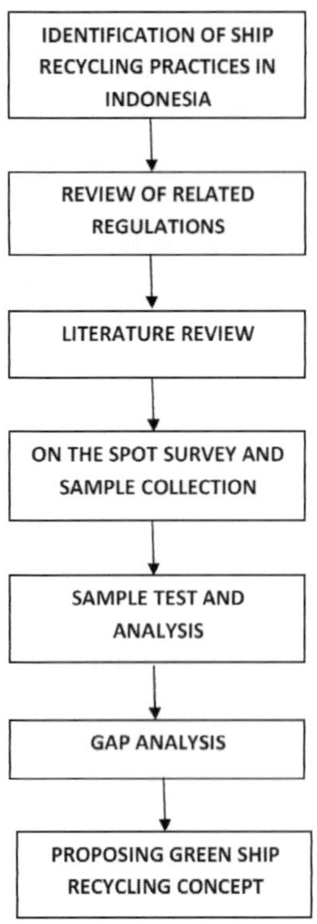

Figure 1 Research methodology

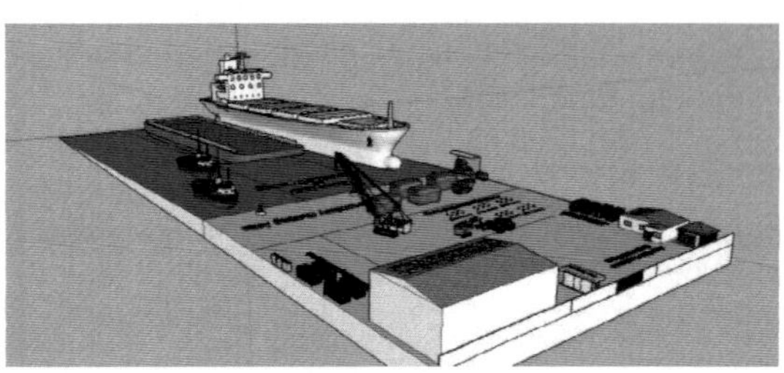

Figure 1 Illustration of air bag green ship recycling concepts

A MODEL SCALE OF UNSINKABLE SMALL PASSENGER SPEED-BOAT

W R Hetharia, E R de Fretes, F Gaspersz and Y Louhenapessy, Faculty of Engineering, Pattimura University, Ambon, Indonesia.

SUMMARY

An intensive study was executed on the small passenger boats due to the accidents occurred in the archipelago regions. A concept of preventing the incoming water on board and additional buoyancy forces was implemented by the application of solid boxes and air bags. A scale boat model was developed to compute and analyse the model parameters and to construct the model. The boat model was arranged for the details of weight components and their centres. Six scenarios of boat configurations were executed to find the amount of solid boxes and air bags. At the sixth scenario of test, the amount of solid boxes of 18.95 dm^3, inside air bags of 7.99 dm^3 and outside airbags of 4.46 dm^3 were applied to the model. There was an additional weight displacement of 8.23 kg that can float and keep the model on stable condition. The results of model test will be applied for full-scale boat.

Keywords: solid boxes, air bags, unsinkable

NOMENCLATURE

[Symbol]	[Definition] [(unit)]
Δ	Weight displacement (kg)
λ	Scale factor
ρ	Density of water (kg m^{-3})
B	Beam (m)
C_B	Block coefficient
DWT	Deadweight (kg)
GM_T	Initial metacentre (m)
H	Height (m)
L	Length (m)
LWT	Lightweight (kg)
T	Draft (m)
T_M	Draft model (cm)
H_M	Height model (cm)
V_S	Speed (knot)
W	Total boat weight (kg)

1. INTRODUCTION

Transportation of passenger with small fast boats is a special issue that should be considered particularly in the archipelago regions such as in the East Indonesian regions. The fast passenger transport in these regions is highly required due to the need of fast transport between the small islands. In fact, there are a lot of accidents due to some factors occur during the operation of such small passenger boats. Most boats capsize or sink and these conditions cause the loss of life and boats at sea [1, 2, 3]. A previous study by the authors, with small tuna long-boats, [4, 5] produced some prototype of unsinkable boats. The boats were provided by light solid boxes in order to prevent the excessive incoming sea water on board. Therefore, with the reserve buoyancy, the boat will be able to float.

The concept of unsinkable tuna long-boats, with the hull material of Fibreglass Reinforced Plastic (FRP), was applied to the small fast passenger boats. This concept was developed to prevent the passengers and boats from sinking. However, since the internal spaces on board are required for the passenger then the application of solid boxes will be critical. Early computations showed that the amount of solid boxes on board will not ensure the boat to float due to the excessive incoming water. Therefore, the boat, with hull material of FRP, will be provided also with additional air bags installed on board and outside the boat. The air bags are placed at small boxes on board and will be expanded as required. The air bags are connected to the compressed air tank stored inside the passenger seat box. At the same time, some side floaters (air bags) were installed at the outside of the boat to provide additional buoyancy and stability capability to prevent the boat from sinking and capsizing. The data from the existing full-scale boats were selected as a reference to construct the model and full-scale boat.

At the first step of the study, this concept was applied to a scale model boat. A scale factor, λ = length of the boat/length of the model = 4 was selected to develop the model. All details of model such as dimensions, geometrical hull form and weight components were adjusted to the scale model. The model and its components on board was tested at the tank. The results of the test were estimated for the full-scale boat.

2. LITERATURE REVIEW

2.1 SAFETY ASPECT OF SMALL BOAT

The contribution of small fast passenger boats in the archipelago regions is required due to less travel time at sea. Since the input passenger is small in the small islands then, the operation of several small boats is reasonable compared to one big boat. There are a lot of boat accidents due to some factors such as bad sea conditions, strong wind, boat configuration and stability, overload, safety and communication equipment, low skill of boat operators and human errors [2, 3]. In this case, the control of boat operation from the authority boards should be in high priority to reduce the accidents at sea.

The concept of preventing the small boats from sinking by previous study from the authors has been applied for local boat builders to construct the unsinkable tuna long boats. This concept as mentioned above will be applied for small fast passenger boats.

2.2 BASIC CONCEPT OF SHIP FLOATING

Basic principal of a ship to be float is the weight displacement of the ship equals to the total weight of ship [6, 7, and 8]. This is showed in the equation (1) as follows:

$$\Delta = W \quad (1)$$

where: Δ = Weight displacement
W = Total weight

The total weight of a ship consists of lightweight (LWT) and deadweight (DWT) [9, 10, 11, 12]:

$$W = LWT + DWT \quad (2)$$

The component LWT consists of boat weight, engine and equipment, while the component DWT consists of passengers, boat operators, luggage and fuel.

Weight displacement of ship (Δ) may be computed as:

$$\Delta = C_B \times L \times B \times T \times \rho \quad (3)$$

where: L = ship length B = ship beam
T = ship draft C_B = block coefficient
ρ = specific weight (1.025 ton/m^3, sea water)

Therefore, for a ship to be float at the draft (T), then the total boat weight (W) equals to weight displacement (Δ).

For a case when a ship to be sink at ship height (H):

$$W_{total} > \Delta = C_B \times L \times B \times H \times \rho \quad (4)$$

In this case the component LWT is constant then the additional weights are coming from the additional DWT which is the incoming water on board.

2.3 AN INNOVATION OF UNSINKABLE BOAT

In order to reduce the incoming sea water some materials should be provided on board. As it is showed in Figure 1, the inside boat are composed with volume of boat structures and solid boxes. With this composition, there will be the amount of sea water fill the boat. The excessive of sea water may cause the boat will sink. To prevent the boat from sinking then there should be applied sufficient air bags on board. At this condition, the boat may still because of limitation of air bags. The limitation of air bags is due to the spaces required for the passengers.

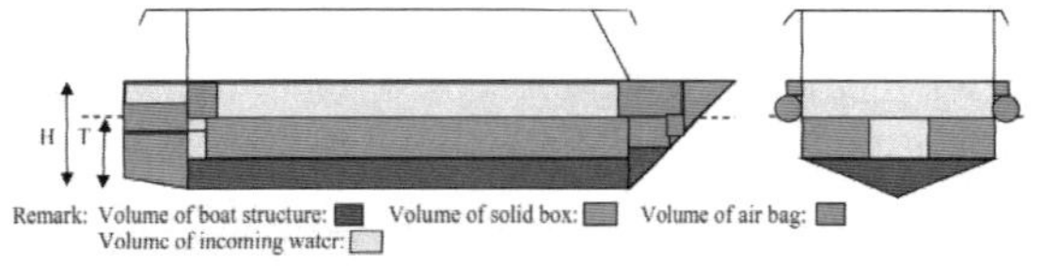

Figure 1: Composition of solid box and air bag on board

In addition, at this configuration, the boat will loss stability due to the small metacentric height. Therefore, an additional air bags should be provided at the outside boat. The application of outside air bags will increase the buoyancy and stability performance of the boat.

In this study, the volume of boat structures are covered by a sheet of FRP material which acts as double bottom and double skin. The solid boxes are composed by two unit of passenger seats and closed rooms placed at the aft and fore parts of the boat. The volume of air bags come from expanded bags installed on board and outsides of the boat. The bags are connected to a unit of compressed air tank installed at inside seat. In normal conditions, the bags are stored in the small boxes on board and outside the boat.

3. METHODOLOGY

3.1 COLLECTING THE BOAT DATA

The boat data are collected in order to develop the boat prototype and its model. In this case, the boat parameters are computed and investigated in order to find the amount of required solid boxes and air bags. The boat data were collected at some local ports in Ambon Island (Figure 2).

Figure 2: Boat data collecting

A model test was provided in order to confirm the results of computation before applying to a full-scale boat. A scale factor was determined λ = length of full-scale boat/length of model = 4. The relationship between the full-scale boat and the model are presented at Table 1.

Table 1. The Dimensions of Full-Scale Boat and Model

No	Boat parameters	Symbol	Value		Unit
			Full-Scale	Model	
1	Length Overall	L_{OA}	6.61	1.652	m
2	Length of Waterline	L_{WL}	6.35	1.587	m
3	Beam	B	1.40	0.350	m
4	Draft	T	0.40	0.100	m
5	Height	H	0.65	0.163	m
6	Speed	Vs	13.0	-	knot

Other boat data are described as follows:

Passenger: 15 persons Operator: 1 person
Hull Material: FRP Autonomy: up to 30 n.m
Luggage: @ 20 kg Fuel: 40 l
Prime mover: 40 HP (outboard)
Equipment: anchor, fuel tank

3.2 DEVELOPING BOAT DRAWINGS

The boat data were developed to obtain the geometrical hull form in order to construct the model and full-scale boat. In addition, computation of boat weight and the components of spaces on board will be verified at the real model and full-scale boat. The lines plan and general arrangement of the boat are presented in Figures 3 and 4.

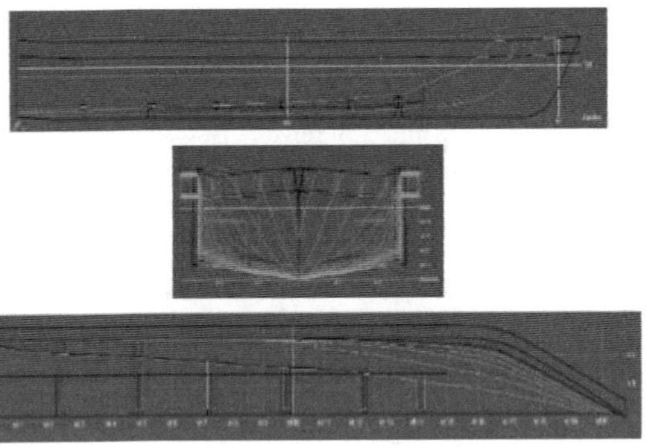

Figure 3: The lines plan

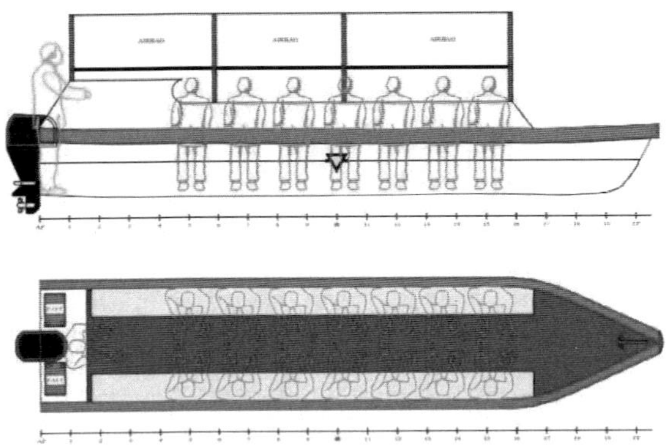

Figure 4: General arrangement

3.3 DEVELOPING AND TESTING THE MODEL

A model was developed in order to confirm the result of computation. The dimensions and geometrical hull form of the model was adjusted to a scale factor $\lambda = 4$. The model hull geometrical was developed based on the lines plan of the full-scale boat. The model was constructed in three stages which were inner template, model template and boat model. The solid boxes, equipment and all testing procedures were performed during constructing the model. The model was built and tested at the Laboratory of Ship Construction, Faculty of Engineering.

Figure 5: The construction of boat model and air boxes

4. RESULTS AND DISCUSSIONS

4.1 RESULTS

The results of model test are presented in Table 2. There are several scenarios during the test. For example, at the scenario number 1, the model was provided with solid boxes on board. The model was set to be float at designed waterline due to the design loads (passengers, luggage and fuel) applied to the model. The result was the model floats at the designed waterline.

At scenario number 2, the model was provided with solid boxes. All loads and incoming water filled the model. As a result, the model was sink. At scenario number 3, the model was provided with solid boxes and air bags. All loads and incoming water filled the model. As a result, the model was still sink. At scenarios number 4, 5 and 6, the model was provided as similar to the scenario number 3 but there was additional side floaters 1, 2 and 3 respectively, with different configurations.

At the scenarios number four to six, the model was provided with additional side floaters installed at outside the model, under the fenders. The side floaters contribute to additional buoyancy as well as stability parameter. The purpose of scenario is to determine at which configuration the model will be float.

The volume displacement of the model (including fender) is 58.74 dm^3 and weight displacement are 58.73 kg. Total volume of solid boxes are 18.95 dm^3, passenger, luggage and equipment are 7.51 dm^3 and air bags inside the model are 7.99 dm^3. The total incoming water at scenario number 2 are 29.81 dm^3 and for scenario number 3 to 6 are 24 dm^3. The additional displacement volume for floater number 1, 2 and 3 are 1.11 dm^3, 2.51 dm^3 and 4.46 dm^3 respectively.

The model test is showed in Figure 6 (for the scenario number 6). In this scenario, the configuration of the model consist of inside solid boxes and air bags, passengers and equipment, side floaters and incoming water. The result of this test was the model floats.

Figure 6: Model testing at the tank

Table 2. Weight components of boat and model

No	Model Parameters	Value	Unit
A	**Float to designed waterline (T_M = 10 cm)**		
	Scenario 1: Model + solid boxes + passengers		
1	Model weight	35.68	kg
2	Volume displacement	34.94	dm^3
3	Weight displacement	35.81	kg
4	Difference	0.40	%
	Testing Result: Model afloat at draft		
B	**Load to model height (H_M = 163 cm)**		
5	Volume displacement model	56.44	dm^3
6	Volume displacement fender	2,29	dm^3
7	Volume displacement model & fender	58.73	dm^3
8	Weight displacement model & fender	58.73	kg
	Scenario 2: Model + solid boxes + passengers + water		
9	Model weight	35.68	kg
10	Weight of water on board	29.81	kg
11	Total weight model + water	65.49	kg
12	Weight displacement	58.73	kg
13	Weight displacement – total weight	-6.76	kg
	Testing Result: Model sinks		
	Scenario 3: Model + solid boxes + air bags + passengers + water		
14	Model weight	35.68	kg
15	Weight of water on board	24.00	kg
16	Total weight model + water	59.68	kg
17	Weight displacement	58.73	kg
18	Weight displacement – total weight	-0.95	kg
19	Initial metacentre GM_T	3.75	cm
	Testing Result: Model sinks		
	Scenario 4: Model + solid boxes + air bags + side floater 1 + passengers + water		
20	Model weight	35.68	kg
21	Weight of water on board	24.00	kg
22	Total weight model + water + floater 1	60.42	kg
23	Weight displacement	59.84	kg
24	Weight displacement – total weight	-0.58	kg
25	Initial metacentre GM_T	3.53	cm
	Testing Result: Model sinks		
	Scenario 5: Model + solid boxes + air bags + side floater 2 + passengers + water		
26	Model weight	35.68	kg
27	Weight of water on board	24.00	kg
28	Total weight model + water + floater 1	61.28	kg
29	Weight displacement	61.24	kg
30	Weight displacement – total weight	-0.04	kg
31	Initial metacentre GM_T	3.30	cm
	Testing Result: Model sinks		
	Scenario 6: Model + solid boxes + air bags + side floater 3 + passengers + water		
32	Model weight	35.68	kg
33	Weight of water on board	24.00	kg
34	Total weight model + water + floater 1	61.72	kg
35	Weight displacement	69.95	kg
36	Weight displacement – total weight	8.23	kg
37	Initial metacentre GM_T	3.10	cm
	Testing Result: Model floats		

4.2 DISCUSSIONS

It is seen from the results presented in the Table 2 that in the scenario number 2 to 5 the model is sink. This is because the total weight of the model exceed the displacement weight. Meanwhile, at the scenario number 6, the total weight is less than the displacement weight.

In addition, there are some values of metacentre height provided for the scenario number 3 to number 5. However, since the model was fully sink then the contribution of side floaters for the stability was not useful.

As it is showed in the scenario number 6 that there is a reserve weight displacement that makes the model still afloat as showed in Figure 6. In this case, the model needs:

- Volume of solid boxes : 18.95 dm^3
- Volume of inside air bags : 7.99 dm^3
- Volume of outside air bags: 4.46 dm^3.

5. CONCLUSIONS AND FUTURE STUDY

5.1 CONCLUSIONS

Some points may be concluded from the results of work which are:

- ➢ The model will float when the weight displacement is greater than the total weight of the model.
- ➢ The solid boxes and air bags installed on board contribute to reduce the incoming water.
- ➢ The air bags installed outside the boat give additional displacement weight to float the boat.
- ➢ The configuration of the boat evaluated in this study needs a total weight displacement of 69.95 kg and the total solid boxes and air bags of 31.4 dm^3 to float the boat model.

5.2 FUTURE STUDY

It needs an additional future study to prove the result for the full-scale boat. The study is being executed now by constructing the full-scale boat. The boat will be tested at sea to prove the results of computation and the results from the model test. The scale factor λ = 4 will be applied to extrapolate the results from the model test to the full-scale boat.

6. ACKNOWLEDGEMENTS

Special thank is given to The Ministry of Research, Technology and Higher Education, The Republic of Indonesia for the research grant of "Penelitian Terapan Unggulan Perguruan Tinggi" 2018 - 2019 to execute this research

7. REFERENCES

1. ANONIMOUS, 'Data Kecelakaan Kapal Laut', *BASARNAS Ambon*, 2016

2. HETHARIA, W. R., de FRETES, E. R., GASPERSZ, F., and TIANOTAK, A. J., 'Kajian Tentang Beberapa Aktifitas Masyarakat Maluku di Sektor Kelautan', *Prosiding Kongres Budaya Maluku*, Ambon, 2014

3. HETHARIA, W. R., GASPERSZ, F, FENINLAMBIR, A., 'Evaluasi Parameter Stabilitas Kapal Penumpang Kecil', *Prosiding Seminar Nasional Fakultas Teknik Universitas Pattimura Ambon,* 26 April 2018, ISSN : 2620-3995, 2018.

4. HETHARIA, W. R., de LIMA, E. J, 'Design and Evaluating of Unsinkable Tuna Long Boats'. *Proceeding of The 6th Regional Conference on Marine Technology (MARTEC 2008)*. University of Indonesia, Jakarta 26 – 27 August, 2008.

5. HETHARIA, W. R., 'Design and Application of Unsinkable Tuna Longboat for Local Fishermen', *OCCASIONAL PAPERS No. 58 (March 2017)*, Kagoshima University, Japan, 2017

6. BARRAS, B. and DERRETT, D. R., 'Ship Stability for Masters and Mates', *Butterworth – Heinemann, 6th edition, An Imprint of Elsevier Ltd.*, Linacre House, Jordan Hill, Oxford OX2 8DP, 2006.

7. BIRAN, A. B., 'Ship Hysrostatics and Stability', *Technion – Faculty of Mechanical Engineering, 1st publication, Butterworth – Heinemann, An Imprint of Elsevier Ltd.*, Linacre House, Jordan Hill, Oxford OX2 8DP, 2003.

8. MOORE, C. S., 'Intact Stability – The Principles of Naval Architecture Series', *The Society of Naval Architects, SNAME*, Jersey City, New Jersey, 2010.

9. PARSONS, M. G., 2003, 'Parametric Design', Ship Design and Construction - *Chapter 11, Volume. 2., SNAME Publication*, Jersey City, NJ, USA, 2003.

10. WATSON, D. G. M., 'Practical Ship Design', *Elsevier Ocean Engineering Book Series*, Volume I, ELSEVIER, 1998.

11. LEWIS, E. V., 'Principles of Naval Architecture, Stability and Strength', *SNAME Publication, Vol. I*, Jersey City, New Jersey. 1988

12. Van DOKKUM, K, 'Ship Knowledge - A Moderen Encyclopedia', *DOKMAR*, Enkhuizen, The Netherland, 2003

8. AUTHORS BIOGRAPHY

Wolter Roberth Hetharia, Currently holds the position of lecturer at Faculty of Engineering, Department of Naval Architecture, Pattimura University. He is responsible for the courses of ship design and resistance. Several papers have been published concerning ship resistance and optimization and fishing boats. He also presented some papers at RINA conference and other international conferences. He and his team executed several research projects funded by Department of Higher Education, Research and Technology.

Eliza R de Fretes is the lecturer at The Faculty of Engineering, Department of Naval Architecture, Pattimura University. He is responsible for the course of ship hydrodynamic and stability. Several papers have been published concerning ship hydrodynamic and stability.

Fella Gaspersz is the lecturer at The Faculty of Engineering, Department of Naval Architecture, Pattimura University. She is responsible for the course of ship design. Several papers have been published concerning ship design.

Yandri Louhenapessy is the lecturer at The Faculty of Engineering, Department of Mechanical Engineering, Pattimura University. He is responsible for the course in mechanical fields and a member of research study by the authors.

THE DEVELOPMENT OF INTACT STABILITY CRITERIA: THE WORK ON SMALL SHIP UP TO 24 M OPERATE IN INDONESIAN WATERWAYS

S Rosihan, **S Anggara**, **M Zaky**, **A A Syariful** dan **D R Mauliani**, Research and Development Division - Biro Klasifikasi Indonesia, Indonesia

SUMMARY

Recently, the IMO 2008 Intact Stability (IS) Code has been widely used to ensure the safe operation of the ship by measuring ship stability level. This code contains stability criteria that applicable to the vessels with 24 m in length and above. The criteria has been adopted by Flag Administrator and as well as its Recognized Organization (RO). This 2008 IS code was developed since 1993, however, in some cases, the ships less than 24 meter (small ship) meet that her stability level are not complied with IS code stability criteria. It is assumed that is caused by differences of hull shape compared to typical large ship, and these make small ship more sensitive to its stability and seakeeping. Furthermore, it is needed to clarify the phenomena on the small ship. In this work, the seakeeping characteristics of small ship and its related stability level will analyze. The number of small ship data are compiled and used in the numerical analysis to derive the stability level from its seakeeping performance. Thus, the results are compared with stability criteria of the reference's requirements.

Keywords : *Stability, Criteria, Small Vessel*

NOMENCLATURE

B	Breadth (m)
D	Depth (m)
f	Freeboard (m)
GM	Center of Gravity to Metacenter (m)
GZ	Righting arm (m)
Hs	Significant wave height (m)
LBP	Length between Perpendicular (m)
T	Draft (m)
Tp	Wave peak period (s)
Tr	Ship roll natural period (s)
VCG	Vertical distance of center of gravity (m)

1. INTRODUCTION

Stability may defined as a measure of the vessel's ability to get back on an even keel after having suffered a heel. It is determined by the characteristics of the vessel, such as hull form and weight distribution and how the vessel is operated. The stability of a small vessel is not a constant condition since they are very sensitive to environment load such as wind and wave. Thus ship stability is one of primary parameter to be concerned since in design stage.

The safety problem of small vessels is a major issue across the world. Each year there is an average of 24,000 fatalities and 24 million non-fatal accidents [1]. In the period 2011-2015, almost 1,368 small vessels have been involved in 4,620 maritime accidents. 66% of total accident was counted as capsizing criteria that stability take over. This huge number accident lead to deep concern for maritime organization to develop specific standard for small vessel.

Number of criteria provided by ISO 12217-1 set some requirement which more relating to operating procedure rather than design engineering [2]. Meanwhile IMO/FAO gives more clear design standard for minimum stability performance [3]. Some flag administration has also develop their independent criteria instead of adopting from IMO. Other criteria may be imposed for novel design. Throughout the development of those code, in view of a wide variety of types, sizes of ships and their operating and environmental conditions, problems of safety against accidents related to stability have generally not yet been solved. Furthermore, design technology for modern ships is rapidly evolving and the Code should not remain static but be re-evaluated and revised, as necessary.

This study is working on development of small vessel stability criteria based on typical of the small vessel operating Indonesia and respect to environmental of Indonesia seaway. The result is compared to referenced criteria in order to know the applicability.

2. INTACT STABILITY CRITERIA

2.1 GENERAL CRITERIA

To ensure safety stability, some criteria has been introduced. The IMO (International Maritime Organization) has issued minimum stability criteria for different types of vessel and these criteria are taken into account at the vessel's design stage. IS Code 2008 criteria, also recalled on IMO res. A.749 (18), which give minimum requirement of ship with 24m length and over is commonly adopted by flag administrator and/or Recognized Organization (RO) [4]

Table 1. Model main dimension

Para-meter	REDE	Ferry Boat	Ferry roro	Patrol	Tug boat	Crew boat	Tug boat	Tug boat	Tug boat	Tug boat	Tanker	FiFi Boat
LBP	19.84	10.30	18.95	12.60	21.80	11.08	23.50	24.50	8.00	17.26	15.57	15.089
B	6.30	3.35	8.00	3.01	7.50	3.00	7.30	7.74	3.00	5.366	4.2	3.346
D	2.20	1.69	2.70	2.75	3.50	1.45	3.50	3.76	1.50	2.15	2	1.83
Draft	1.50	0.54	1.96	0.63	2.86	0.59	3.01	2.85	0.99	1.378	1.592	1.04

The code was developed based on stability of passenger and general cargo ship under 100m length (res.167 (ES.IV)) and ship carrying deck cargo (res.206 (VII)) [5]. Later the application of code then expanded to fishing vessel, special purpose ships, offshore supply vessels, mobile offshore drilling units, pontoons, dynamically supported craft and container ships.

IS code 2008 General criteria is break down as follow [4]:

a). The area under the righting lever curve (GZ curve) shall not be less than 0.055 meter-radians up to fig = 30° angle of heel and not less than 0.09 meter-radians up to fig = 40° or the angle of down-flooding if this angle is less than 40°. Additionally, the area under the righting lever curve (GZ curve) between the angles of heel of 30° and 40° or between 30° and, if this angle is less than 40°, shall not be less than 0.03 meter-radians.

b). The righting lever GZ shall be at least 0.2 m at an angle of heel equal to or greater than 30°.

c). The maximum righting lever shall occur at an angle of heel not less than 25°. If this is not practicable, alternative criteria, based on an equivalent level of safety see footnote, may be applied subject to the approval of the Administration.

d). The initial metacentric height GM0 shall not be less than 0.15 m.

2.2. STABILITY CRITERIA FOR SMALL SHIP

Since decades, Several Flag administration e.g. Malta [6] has used IS Code General Criteria for small yacht. The criteria has also been being used by Indonesian RO for reviewing stability all type and all size of ship, including small ship with length under 24m. This may cause ship not able to comply this criteria due to overestimated requirement. As known, hull shape of small ship is quite different compared to typical large ship, i.e. tug boat vs general cargo ship, the stability and seakeeping characteristic shall not be likened.

Therefore specific criteria for small ship need to be set up. Today the only reference provided by IMO is res. A.207 (VII) [7] which cover fishing vessel up to 30 meters of length. This code is applied in lieu of IMO res. A.168 (ES.IV) [8] if there is no enough data provided by owner. IMO/FAO/ILO [3] also adapts this recommendation. The criteria may be used for several type of ship but limited in similar hull shape. ISO 12217-1 [2] is more widely applicable, but specific for only ship with length of 6 to 12 m. The standard aim to obtain safety operation in addition of design compliment. Criteria consist of recommendation for minimum height and angle of downflooding point, maximum righting arm, and area under GZ curve.

Meanwhile, number of flag state has taken the initiation in developing their own criteria which based on register data of ship operating in the area, e.g. Sweden, Russia and FRG. In this case, metacentric height is take into account to measured stability level [9].

IMO res. A.207 (VII)

$$GM \ge 0.53 + 2B \cdot \begin{pmatrix} 0.075 - 0.37\dfrac{f}{B} + 0.82\left(\dfrac{f}{B}\right)^2 - 0.014\dfrac{B}{D} \\ - 0.032 \cdot \dfrac{\lambda_S}{L} \end{pmatrix} \cdot B$$

IMO/FAO/ILO

- Design Category A (Hs > 4m) and B (2m < Hs ≤ 4m)

$$GM \ge 0.117B \cdot \left(\frac{B}{D} - 2.2\right) + \left(1.773 \cdot \left(\frac{T}{D}\right)^2 - 2.646\frac{T}{D} + 1.016\right) \cdot B$$

- Design Category C (0.3m < Hs ≤ 2m) and D (Hs < 0.3m)

$$GM \ge 0.059B \cdot \left(\frac{B}{D} - 2.2\right) + \left(2.085 \cdot \left(\frac{T}{D}\right)^2 - 2.857\frac{T}{D} + 0.99\right) \cdot B$$

- Sweden $GM \ge 0.41 + 0.014\frac{\cdot B}{f}$
- Russia $GM \ge T \cdot \left(0.34 \cdot \left(\frac{B}{T} - \frac{f}{B}\right) - 0.47\right)$
- FRG $GM \ge 0.5 + 0.035\frac{\cdot B}{f}$

All these seems not enough since development is still evolving up today.

The development of small ship stability criteria is absolutely necessary in state where portion of fleet under 24 m is promising, e.g. Indonesia. The primary issue faced today is that number of ship are not registered to flag state, the reason is due to lack of stability compliment. In fact, the ship has been operating safely so far without any obstacle. This case become main concern in engineering point of view. Therefore, through this study, writer has successfully established the small ship stability criteria based on RO database that may later be used for general type of ship.

3. MODEL

Model selection was taken proportionally to ship type distribution registered in one of Indonesian RO database, so called IRO. They consists of almost all type of ship in it class e.g. Tug boat, ferry, tanker, patrol boat, crew boat, fifi boat. The extracted data may be described as follow:

- The length variates in range of 8 m to 24.5 m. See Table 1
- GT show exponential distribution as LBP function and linear distribution as function of hull volume
- L/B ratio is ranging from 2.7 – 3.2, while B/D ratio is from 1.5 to 3.0. See Figure 1 and 2.

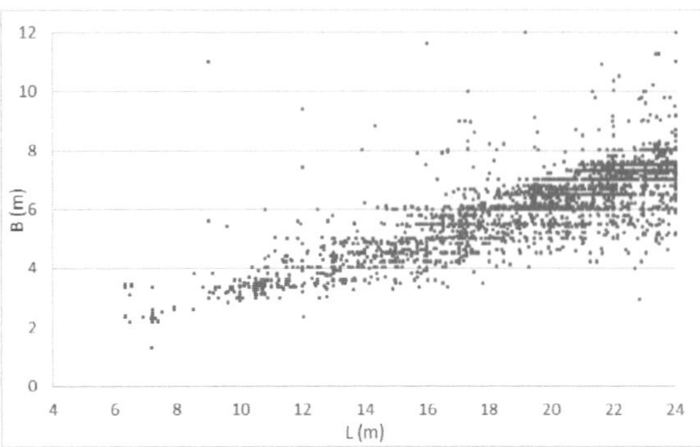

Figure 1. Distribution of L/B ratio small ship registered in IRO

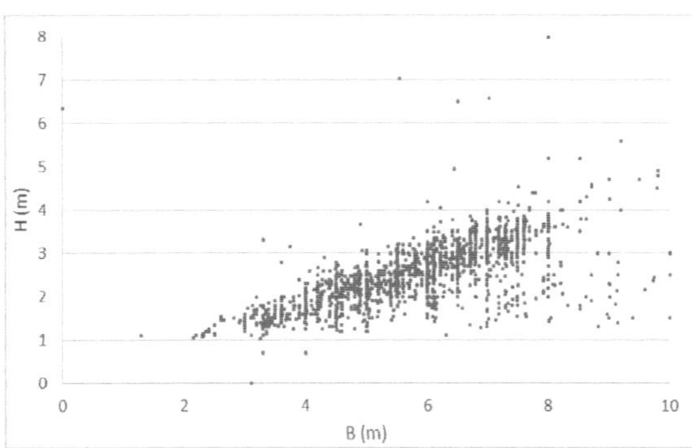

Figure 2. Distribution of B/D ratio small ship registered in IRO

In order to comprehensively cover all range of dimension of common small ship in data base, the sampling then variates as per L/B 2.75; 3.00; 3.25; 3.5 and 3.75. Total of 60 model created with displacement scaled, thus draft is constant for each type of ship. But VCG to D is set variating as per D/T. The aim is to capture enough variating behavior from the sampling. The analysis was done with ship considered in maximum operation draft and in static condition. Previously, created model for sampling has been verified to minimize gab to the actual ship. Error shall be less than 5% for displacement and 1% for other parameter.

Stability characteristics through its parameters such as initial GM are examined to then be used as a typical reference criteria for small ship stability. Reported GM is allowable value respect to IS Code. The reason of GM to become sought parameter is the fact that it is simple to calculate without any other supported document.

4. SEAKEEPING ANALYSIS

In particular, the safety of a ship in a seaway involves complex hydrodynamic phenomena which up to now have not been adequately investigated and understood. Static analysis is considered not enough to examine ship performance in seaway. Because in fact, any ship in wave will be exhibiting large righting lever variations between wave trough and wave crest condition, this may lead to the worst case. In addition, principally, the static analysis is derived from seakeeping analysis.

In dynamic stability, roll radius gyration is working like GZ which form restoring moment. Beside of transverse mass distribution, this parameter is more influenced by mass vertical position. Furthermore, the ship's behavior also depend on environmental condition of seaway where ship operated. Wind and wave become main factor to take into account. The energy of those variable statistically can be expressed with probability density function through Rayleigh distribution, so called as spectrum. Irregular wave and wind can be converted to superposition of number of regular sinusoidal form. In this study, seakeeping analysis was also performed to have detail sight of small ship behavior. Hs 1.5 m and Tp 6.4 s is used to represent average condition of Indonesia waterway [10]. The value is equal to Beaufort 4 and hereafter transformed to Pierson Moskowitz spectrum.

Roll radius gyration is calculated based on IMO res. A 562(14) [11] which is adopted by IS code weather criteria. Simulation has been done with all direction wave heading and 45 degree spacing. Ship speed was adapted to operational condition. Hereafter, maximum response in roll direction is investigated by using strip theory and assumed as maximum heeling of the ship.

5. DISCUSSION

Initial investigation was done to see how much small vessel unable to comply to IS code criteria. Each sampling has been examined to capture allowable GM respect to IS code. Then, the results was compared to actual GM and it was proven that 25% from total sampling failed to comply (see Figure 3). This is similar to finding of [12] in where about 50% from 4997 unit of small ship are fail to comply to IS code.

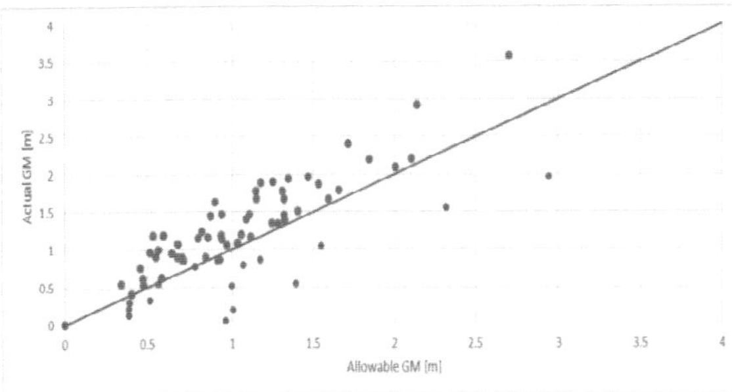

Figure 3. Allowable GM respect to IS Code

Further sight show that the major sampling fail to comply criteria 1 to 5 refer to [13], except for GM criteria which more sampling capable to comply with. It concludes that those criteria, other than GM, are more conservative, and in other word there is more space to modify GM requirement as compensate other criteria in IS Code.

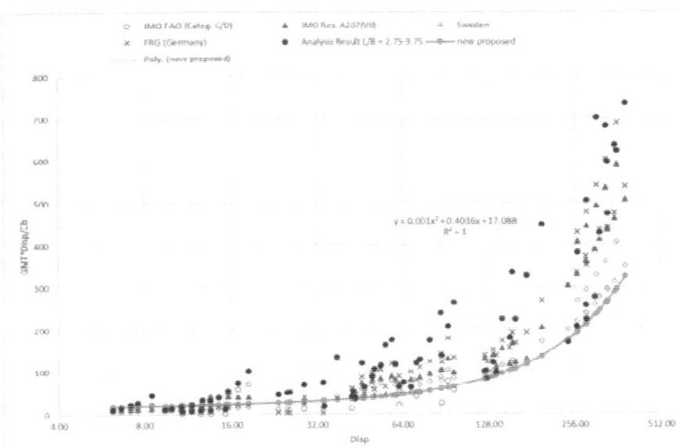

Figure 4. GM distribution

Figure 4 shows that GM as function of displacement forms exponential distribution. As displacement increase the need of GM is dramatically increased. Regression formula of the data may be expressed as follow:

$$GM = \left(\frac{\Delta}{1000} + 0.4 \right) Cb + \frac{16.5}{LBT}$$

Other references criteria from several country and standard, e.g. IMO res. A207, IMO-FAO (category C and D) and local regulation from flag administration of Swedia and FRG, are also plotted and show similar trend with this study.

Minimum requirement for ship with length over than 12 m can be clearly approached by the formula. However, the formula is slightly conservative if applied to ship under 12 m or having displacement less than 16 tons. In this cases, they may need special consideration. For example, IMO-FAO gives smaller GM value compared to other standard because this code is specifically for ship with length under 12 m.

The interesting issue happen to GM changing pattern when B transforms. With the fixed length and draft, the GM and GZ change to form exponential function due to B transformation. So far, ship length is understand not to affect stability significantly, thus 2 ship with same draft even having different main dimension would principally show the same necessary of GM value [14].

Seakeeping has been performed and response is plotted to figure 5. The used GM was based on proposed formula. The distribution of roll period may be expressed by following linear function.

$$Tr = \left(\frac{0.25}{GM} + 0.38 \right) \cdot B$$

Roll period calculated by IMO res. A 562 and IMO FAO show similar trend but higher value compared to simulation. This may be understood because of different assumption of inputted parameter. The IMO was assumed that ship is in dead condition means non propulsion system works. This may be endangered by resonant roll while drifting freely. Meanwhile proposed formula for roll period has been considering the effect of operational speed and environmental condition of Indonesia waterway. However this study has already been considering wave coming from all heading angle in order to anticipating the extreme condition. Where ships in following and quartering seas may not be able to keep constant course despite maximum steering efforts which may lead to extreme angles of heel.

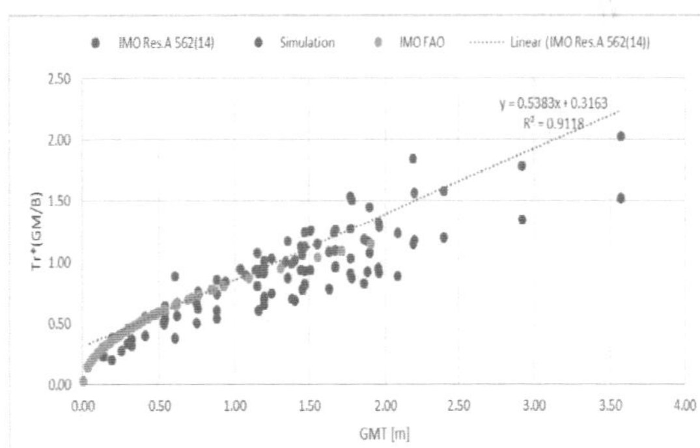

Figure 5. Roll period and GM relationship

6. CONCLUSSION

Having investigation of small ship stability, proposed formula to calculated GM is obtained. This may be used clearly for ship with length 10 – 24 meters. Ship under 10 meter can also refer to the formulas with some consideration since the result slightly conservative. The special study may be separated for the very small ship, since the standard such ISO has also specific concern to such size of ship.

Seakeeping performance has also been studied, the ship's roll period can be approach as well by regression formula which includes Indonesia environmental condition and ship operational speed. It may help in initial design step to have sight of ship roll acceleration. But to be concerned that the response has already taken into account the ship speed to represent the actual operation condition. Meanwhile, Formula by IMO is focused on ship in dead ship situation in order to catch the worst condition due to resonant roll.

7. REFERENCES

[1] Atzampos G., 'A New Era of Fishing Vessel Safety Emerges', *Proceedings of 7th Transport Research Arena,* 2018

[2] International Organization for Standardization, 'Small craft – Stability and buoyancy assessment and categorization', *ISO 12217-1,* 2002

[3] FAO/ILO/IMO; Fisheries and Aquaculture Management Division, ' Safety Recommendations for Decked Fishing Vessels of Less than 12 meters in Length and Undecked Fishing Vessels', 2012

[4] International Maritime Organization, 'Intact Stability General Criteria', *International Code on Intact Stability - Resolution MSC. 267 (85),* 2008

[5] International Maritime Organization, 'Recommendation on intact stability for passenger and cargo ships under 100 meters in length', *Resolution A. 167(ES.IV) adopted to Resolution A.206(VII).* 1971

[6] Malta administration, 'NCV Code for Non-Convention vessels and for Commercial vessels trading exclusively within Maltese Waters', Malta, 2001

[7] International Maritime Organization, 'Recommendation for an Interim Simplified Stability Criterion for Decked Fishing Vessels Under 30 Meters in Length', *Resolution A. 207(VII),* 1971

[8] International Maritime Organization 'Recommendation On Intact Stability Of Fishing Vessels', *Assembly - 4th extraordinary session - Resolution A.168(ES.IV),* 1968

[9] Brown P., 'A Study of Small Ship Stability', *Department of Defense Canberra.* 2000

[10] Kurniawan M.A., Preasetyo F.A., Komaritah S., 'Indonesian Sea state Condition and Its Wave Scatter Map', *MASTIC (Maritime Safety International Conference)* , 2018

[11] International Maritime Organization, 'Recommendation On A Severe Wind And Rolling Criterion (Weather Criterion) For The Intact Stability Of Passenger And Cargo Ships Of 24 Meters In Length And Over', *Resolution A.562(14),* 1985

[12] SLF committee: Republic of Korea. 'Study on application of intact stability criteria to small vessels'. *Revision of the intact stability code: IMO.* 2007

[13] International Maritime Organization. "Mandatory Criteria". *International Code on Intact Stability.* 2008

[14] Kluwe F., 'Development of a Minimum Stability Criterion to Prevent Large Amplitude Roll Motions in Following Seas', *Technische Universität Hamburg*, 2009

APPLICATION OF THE SECOND GENERATION INTACT STABILITY CRITERIA TO AN INDONESIAN RO-RO FERRY SUPPORTED BY MODEL EXPERIMENT

D Paroka, A H Muhammad, S Rahman, Rosmani, M A Azis and **A Yusuf**, Hasanuddin University, Indonesia

SUMMARY

The characteristics geometry of Indonesian ro-ro ferries are different with those used to develope the weather criterion in the intact stability criteria of International Maritime Organization. To evaluate the stability of Indonesian ro-ro ferries using the second generation intact stability criteria, an alternative assessment of weather criterion as the vulnerability criteria level 1 is necessary in order to find consistency of each level of vulnerability criteria. This paper discusses about stability evaluation of an Indonesian ro-ro ferry by using the second generation intact stability criteria with the damping factor correspond to the breadth and draught ratio determined by model experiment. The effective wave slope coefficients as function of vertical centre of gravity are calculated using the formula of weather criterion and the simplified strip theory at the wave frequency the same as the natural frequency of roll. The wind and waves characteristics are based on the scatter wave data of a ferry route in Indonesia. As results, the minimum metacentric height for vulnerability criteria level 1 obtained by model experiment is smaller than that obtained by using the values in the weather criterion. The inconsistency between the vulnnerability criteria level 1 and level 2 appears when the values of parameters in the weather criterion are used. However, it shows the consistency when the damping factor correspond to the breadth and draught ratio in the vulnerability criteria level 1 is determined by model experiment. The obtained safety level for the vulnerability criteria level 2 is 0.00008 correspond to the critical metacentric height 1.3 meters. This result show that the minimum safety level recommended by the International Maritime Organization can be applied to Indonesian ro-ro ferries.

NOMENCLATURE

CI	Capsizing index
C_i	Short-term capsizing probability
H_s	Significant wave height (m)
m_0	Variance deviation of roll motion (rad.^{2})
m_2	Variance of angular velocity of roll (rad.^{2}/s^2)
S_i	Occurence probability of sea state
T_{EXP}	Time exposure (s)
U_w	Mean wind velocity (m/s)
$\Delta\phi_{EA^+}$	Range of residual stability in leeward (rad.)
$\Delta\phi_{EA^-}$	Range of residual stability in windward (rad.)
λ_{EA}	Capsizing rate (1/s)

1. INTRODUCTION

The Indonesian ro-ro ferries are characterized by large large breadth compared to the ship draught to provide a large area of vehicles deck and the shallow water in the port area. In order to easily loading and unloading of vehicles, the freeboard is designed to be smaller around 0.10 of ship breadth [1]. Those ro-ro ferries are used for short inter-island and inland river-seas transportations. The vertical center of gravity could be higher than the ship height because the vehicles deck is located in the main deck and the passenger accommodation is above the vehicles deck. The safety of ship in seaways correspond to stability is evaluated by using the general intact stability criteria and the weather criterion of International Maritime Organization (IMO) [2]. Due to their geometry characteristics are different with those used to develope the criteria, some parameters of the criteria are difficult to comply such as the heel angle with maximum righting arm. This is because the angle of deck edge immersion smaller than 10.0 degrees due to small freeboard and small draught compared to the ship breadth [1]. On the other hand, the initial metacentric height, the area under the righting arm curve as well as the angle of vanishing stability are generally larger compared to other ships types due to large breadth and draught ratio.

The roll-back angle obtained in the weather criterion could be overestimate when it is applied to the Indonesian ro-ro ferries. The breadth and draught ratio of Indonesian ro-ro ferries is generally larger than the maximum ratio in the weather criterion. As result, the damping factor correspond to the breadth and draught ratio in the weather criterion could be overestimate [3], [4]. The effective wave slope coefficient obtained by formula of weather criterion is larger than that obtained by model experiment due to large vertical centre of gravity [4], [5], [6]. The formula to determined the coefficient "C" correspond to the formula of natural roll period could result significant error when it is applied to a ship with large breadth and draught ratio as well as large metacentric height [7]. Therefore, IMO provide an alternatif assessment of weather criterion based on model experiment especially for ships with geometry characteristics different with that used to develope the criteria [8]. The wind pressure used in the weather criterion was for wind velocity of 26 m/s. This wind velocity could be different depend on the location and geography characteristics.

Recently, the second generation intact stability criteria (SGISC) is being in finalization step which is a performance based criteria. The revision of intact stability criteria is meant to consider the ships with geometry characteristics different with those used to develope the previous criteria. The weather criterion has been decided by IMO as the vulnerable criteria level 1 for dead ship condition and the vulnerable criteria level 2 was the probability of ship heel angle exceeding a certain angle such as the downflooding angle or the angle of vanishing stability. The level 1 of the vulnerability criteria means to separate between conventional and unconvetional ships. This criteria opens the challenge to overcome the problem of implementation of stability assessment for ship with geometry characteristics out of the range in the weather criterion. A ship comply with the vulnerability criteria level 1 should also comply with the level 2 of the vulnerability criteria. On the other hand, a ship is found to be vulnerable in the level 2 should also be vulnerable in the level 1. During this finalization, the criteria has been tested by applying to several ships types including large passenger ships as well as ro-pax ships. In some cases, the obtained safety level is inconsistence between the vulnerability criteria level 1 and the vulnerability criteria level 2. Therefore, the standard of minimum safety level especially for level 2 of vulnerability has not yet been decided. The IMO suggested the capsizing index to be 0.04 or 0.06 [9] but in the other document it is proposed to be 0.01. The inconsistency could be occur because the safety level in the weather criterion is calculated for one sea state while in the vulnerability criteria level 2, the casizing index is calculated for all possible sea state based on the scatter wave data of operational area. In the other hand, safety level of a ship could change depending on the sea state of operational area. The discrepancy of safety level between the vulnerability criteria level 1 and the vulnerability criteria level 2 may also occur due to different damping factors used in the weather criterion and the capsizing index especially for ships with breadth and draught ratio larger than 3.5.

This paper discusses about the minimum safety level of ships with geometry characteristics are not covered in the weather criterion based on the SGICS. This is important because most of ships used for inter-island and river-seas transportation especially for short distance have geometry characteristics with large breadth and shallow draught as well as small freeboard. The inconsistency of safety level for dead ship condition mainly occur for ships with lower weather deck and very large breadth means small ratio of freeboard and breadth [10]. Rudakovic and Backalov [11] used the capsizing index of 0.04 as the minimum safety level to analyze operational limitation of a river-sea ship using the SGISC. However they did not analyze the consistency between the vulnerability criteria level 1 and the vulnerability criteria level 2. If the consistency of the criteria can be found, the minimum safety level can be determined in advance and the SGISC can be considered to adopt as stability standard for ships with large breadth and draught ratio as well as very small freeboard. The obtained results can also be used to explain the reason of the consistency when the SGISC applied to such ships geometry characteristics.

2. METHODOLOGY

The second generation intact stability criteria is applied to an Indonesian ro-ro ferry with principal dimension shown in Table 1 and its body plan is shown in Figure 1. The righting arm curve of the ship for full loading condition is shown in Figure 2.

Table 1: Principal dimension of ship

Dimension	Ship (m)
Length perpendiculars (Lpp)	50.50
Breadth (B)	14.00
Height (H)	3.80
Draught (d)	2.70
Metacentric height (GM)	4.23
Vertical centre of gravity (KG)	4.717
Length of bilge keel	18.00
Breadth of bilge keel	0.25

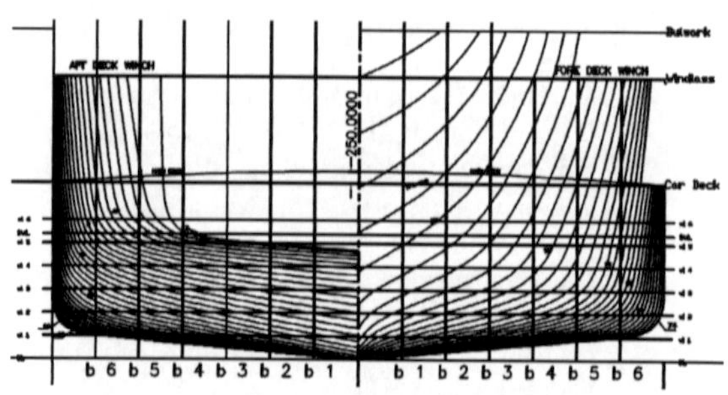

Figure 1: The body plan of subject ship

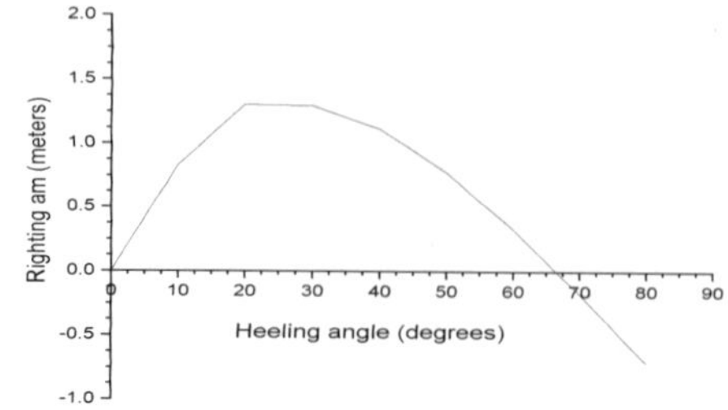

Figure 2: Righting arm curve of the subject ship for full loading condition

The vulnerability criteria level 1 for variation of metacentric height range from 0.50 meters to 8.8 meters are calculated with the damping factor correspond to breadth and draught ratio is determined following the value in the weather criterion and by model experiment. The critical metacentric height is calculated for two different effective wave slope coefficient. The first effective wave slope coefficient is obtained by using the formula of weather criterion [2]. The second one is obtained from the simplified strip theory as recommended by IMO to use in the vulnerability criteria level 2 [12]. Here, the effective wave slope coefficient is determined based on the natural frequency of roll for each vertical centre of gravity. The effective wave slope coefficients obtained by formula of IMO is shown in Figure 3 and those obtained by simplified strip theory for wave frequency the same as the natural frequency of roll as well as model experiment are shown in Figure 4.

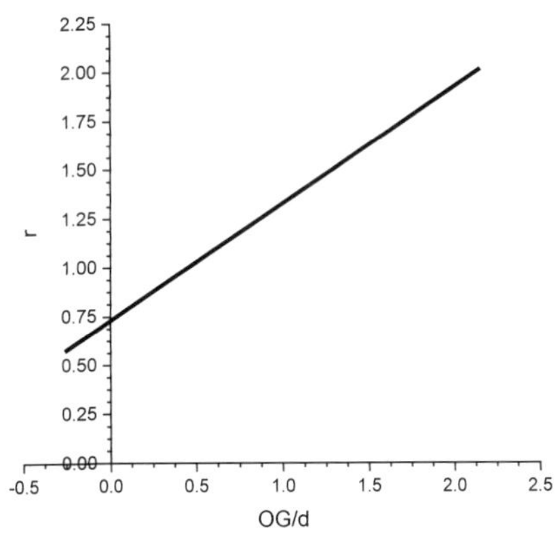

Figure 3: Effective wave slope coefficient obtained by formula of IMO

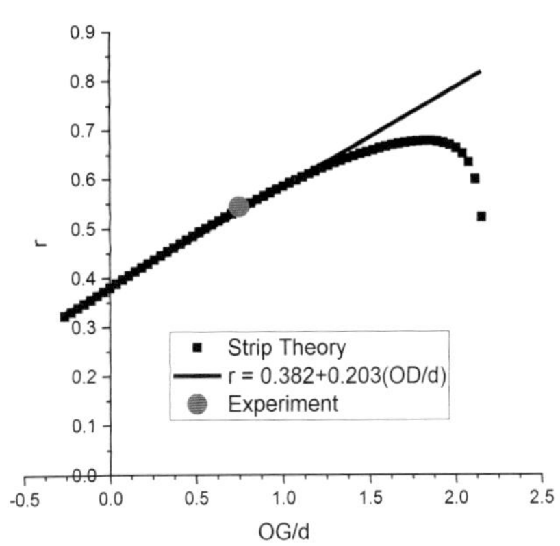

Figure 4: Effective wave slope coefficient obtained by simplified strip theory and model experiment

The effective wave slope coefficient of the ship obtained by formula of IMO is alrger than that obtained by simpified strip theory and by model experiment. For large vertical center of gravity, the effective wave slope coefficient obtained by formula of IMO is larger than 1.0. The result of model experiment is coincide with that obtained by simplified strip theory.

The natural roll period of the subject ship is 4.453 seconds which is smaller than the minimum natural roll period in the weather criterion. Therefore, the wave steepness is determined to be the same as that corresponds to the minimum natural roll period in the criteria. The wind pressure is determined based on the wind speed data of one route of Indonesia ro-ro ferry. Tha maximum wind velocity is 20 m/s correspond to wind pressure of 300 Pa. The wind pressure could be smaller depend on the location. The others variable in the weather criterion formula such as the damping factor correspond to the block coefficient and the bilge keel are determined following the weather criterion.

The capsizing index as the parameter of the vulnerability criteria level 2 for the same range of vertical centre of gravity the same as that in the level 1 is calculated with linear and quadratic damping coefficients determined by model experiment. The effective wave slope coefficients are determined using the same methos as used in the calculation of vulnerability criteria level 1. The statistics wave data of route of ro-ro ferry between Bulukumba and Selayar Island with scatter wave data shown in Figure 5 is used.

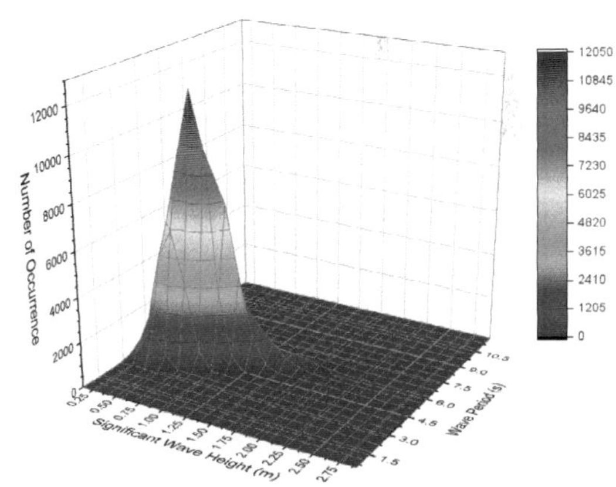

Figure 5: Scatter wave data

The mean wind velocity is calculated as function of significant wave height. This equation is statistically developed based on recorded data of wind and waves with the equation shows as follow [13]:

$$U_w = \left(\frac{H_s}{0.13}\right)^{0.85} \quad (1)$$

This equation is different with formula IMO [12] as the location data colllected is different. The spectrum of wave exciting moment is calculated by using the Jonswap spectrum and the Davenport spectrum is used to calculate the spectrum of wind moment due to wind gustness. The casizing index is calculated using the following equation [9]:

$$CI = \sum_{i=1}^{N} C_i S_i \tag{2}$$

The probability of roll angle exceeding an acceptable angle, here the downflooding angle or the angle of vanishing stability which is the smallest for a certain exposure time is calculated by using the equation as follows [9]:

$$C_i = 1 - exp(-\lambda_{EA} T_{EXP}) \tag{3}$$

Here, the exposure time determined to be one hour. The occurence probability of a sea state is determined as the number of occurence devided by the total number of wave recorded. The wave data recorded shown in Figure 5 is used to determine the probability of each pair of significant wave height and the mean wave period. The capsizing rate in the equation (3) is calculated by using the equation as follows [9]:

$$\lambda_{EA} = \frac{1}{2\pi}\sqrt{\frac{m_2}{m_0}}\left(exp\left(-\frac{\Delta\phi_{EA^+}{}^2}{m_0}\right) + exp\left(-\frac{\Delta\phi_{EA^-}{}^2}{m_0}\right)\right) \tag{4}$$

Here, the residual stability range in leeward and windward directions are determined by maintaining the area under the righting arm curve from the static heel angle due to steady wind to the downflooding angle or the angle of vanishing stability, which one is the smallest. This means that the residual stability range could be larger than the downflooding angle or the angle of vanishing stability depends on characteristics of the righting arm curve. The variance of roll angle and the variance of angular velocity of roll are calculated based on the spectrum of total exciting moment induced by combined action of gustness wind and waves for each sea state.

3. RESULTS AND DISCUSSION

The index "b/a" as the parameter of the vulnerability criteria level 1 for several different vertical centre of gravity are shown in Figure 6. Here, the solid line indicates the results obtained based on the recommended values of each parameter of weather criterion and the dash line show the index "b/a" with the damping factor correspond to the breadth and draught ratio obtained by model experiment and the effective wave slope coefficient calculated by the simplified strip theory. As the damping factor correspond to the breadth and draught ratio obtained by model experiment is smaller than that in the weather criterion, the results obtained by experiment is larger than that obtained based on the recommended values of IMO. The critical metecentric height is 3.9 meters if the values of parameters in the weather criterion is used and that is 1.6 meters when the damping factor correspond to the breadth and draught ratio obtained by model experiment and the effective wave slope coefficient calculated by simplified strip theory are used. The smaller damping factor due to breadth and draught ratio determined by model experiment result in smaller roll-back angle compared to that obtained by using the values in the weather criterion. This smaller roll-back angle is also induced by smaller effective wave slope coefficient obtained by simplied strip theory than that calculated by using the formula in the weather criterion. Therefore, the index "b/a" become large than that by using the values in the weather criterion.

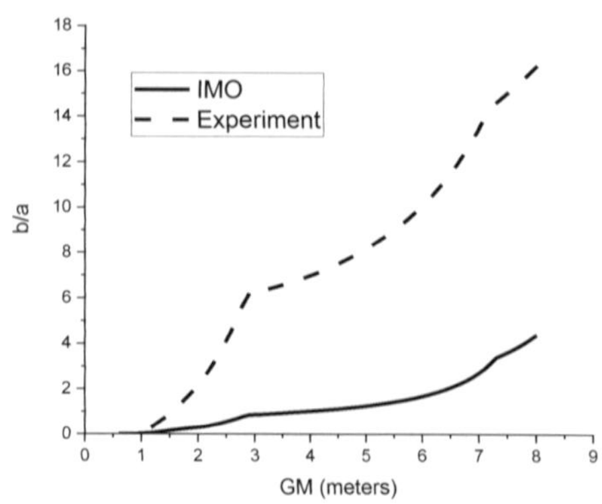

Figure 6: The index "b/a" for variation of matacentric heigh with wind pressure of 300 Pa

Figure 7 show the capsizing index of the subject ship based on the statistics wave data of a route of inter-island transportation in Indonesia serviced by ro-ro ferry. The solid line is the capsizing index with the effective wave slope coefficient obtained by simplified strip theory as function of wave frequency and the dash line for the effective wave slope coefficient calculated by weather criterion formula. Similiar to the vulnerability criteria level 1, the capsizing index by using the effective wave slope coefficient of the weather criterio formula is larger than that obtained by using the simplified strip theory.

This is due to larger effective wave slope coefficient obtained by the formula of weather crirerion.

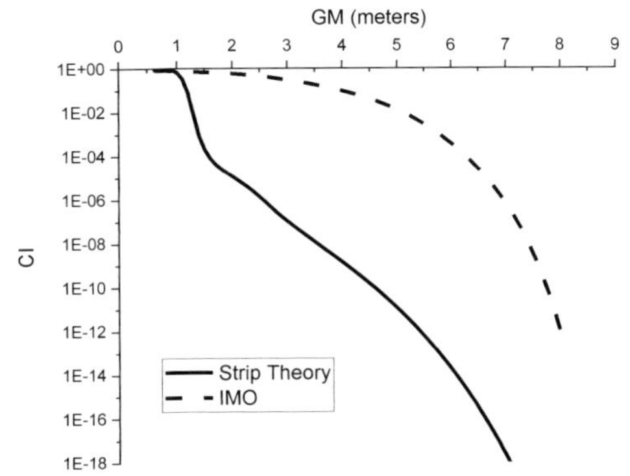

Figure 7: Capsizing index for variation of metacentric height with effective wave slope coefficient obtained by strip theory and formula of IMO.

The vulnerability criteria level 1 and the vulnerability criteria level 2 is not consistence when the effective wave slope coefficient is calculated by using the formula of IMO. The ship is found to be vulnerable in the level 1 when the metacentric height is smaller than 3.9 meters. This metacentric height correspond to a capsizing index of 0.142. This capsizing index is larger than the maximum capsizing index recommended by IMO. If the maximum acceptabel capsizing index follows the IMO recommendation, the minimum metacentric height of the ship is 4.7 meters. This results indicate that the damping factor correspond to the breadth and draught ratio in the weather criterion is overestimate and therefore it should be extended to breadth and draught ratio smaller than 3.50. When the effective wave slope coefficient is calculated by simplified strip theory and the damping factor correspond to the breadth and draught ratio is determined by model experiment, the capsizing index correspond to the vulnerability boundary in the level 1 is 0.00008. This capsizing index is smaller than the maximum capsizing index recommended by IMO. This means that both the vulnerability criteria level 1 and the vulnerability criteria level 2 shows consistency.

Figure 8 and Figure 9 show the relationship between the index "b/a" and the capsizing index for the effective wave slope coefficient calculated by formula of weather criterion and by simplified strip theory, respectively. If the safety standard of IMO with the capsizing index of 0.04 is assumed as the acceptable safety level for the Indonesian ro-ro ferries, the minimum value of index "b/a" should be 1.18. This capsizing index correspond to the metacentric height of the subject ship of 4.7 meters which is the actual metacentric height for full loading condition. This means that applying the values of parameters in the weather criterion to a ship with breadth and draught ratio larger than 3.50 and the ratio between vertical centre of gravity and draught larger than 1.50 result in an overestimate roll-back angle in beam wind and waves so that the vulnerability criteria level 1 and level 2 for dead ship condition becomes inconsistence due to the index "b/a" in the level 1 becomes smaller. The damping moment of the subject ship is larger than that predicted in the weather criterion and the effective wave slope coefficient of the formula of weather criterion is overestimate when it is applied to a ship with large breadth and draugth ratio. This phenomena has been found by others researchers such as Deakin [3] and Sato, et al [14].

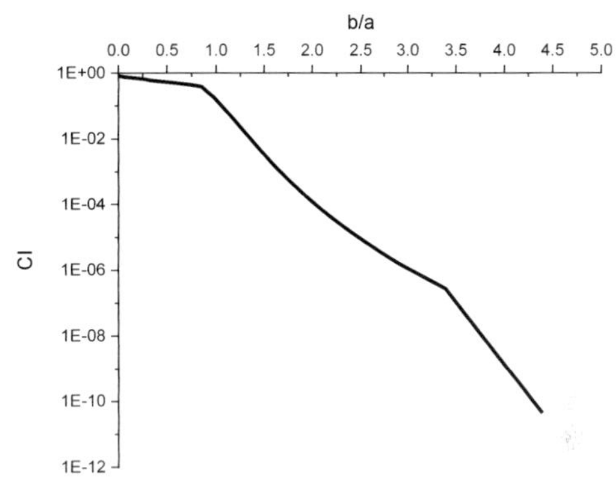

Figure 8: The capsizing index correspond to the index "b/a'' of weather cirterion with damping factor based on weather criterion

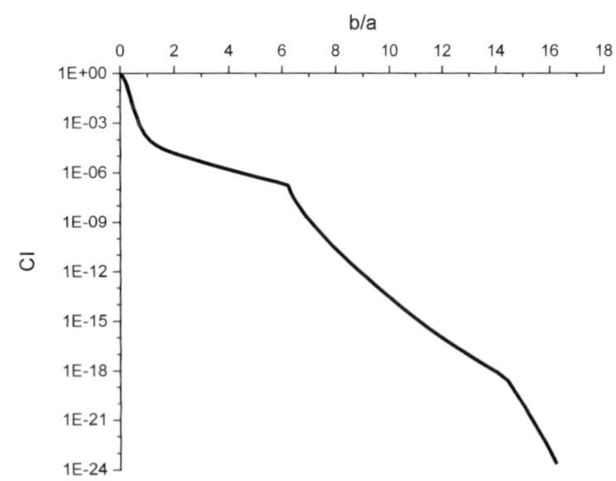

Figure 9: The capsizing index correspond to the index "b/a" of weather criterion with damping factor obtained by model experiment

When the damping factor correspond to the breadth and draught ratio obtained by model experiment and the effective wave slope coefficient obtained by simplified strip theory are used, the capsizing index correspond to the minimum metacentric height obtained in the vulnerability criteria level 1 is smaller than the minimum safety level recommended by IMO. The metacentric height correspond to this minimum safety level is 1.3 meters. The value of index "b/a" in this metacentric height is 0.50 smaller than 1.0 as the safety boundary.

This results show that the same damping factor should be used in both vulnerability criteria level 1 and the vulnerability criteria level 2 in order to find its consistency.

4. CONCLUSIONS

The second generation intact stability criteria has been used to calculate the minimum metacentric height correspond to the vulnerability criteria level 1 and the corresponding capsizing index in the vulnerability criteria level 2. Based on the obtained results and discussions, it can be conluded that the difference method to determine the damping factors in the vulnerability criteria level 1 and level 2 induce inconsistency of the criteria for dead-ship condition. Therefore, it is recommended to used the same estimation method especially for ships with geometry characteristics different with those used to develope the weather criterion. For ship with breadth and draught ratio larger than 3.50, the inconsistency appears due to the damping factor correspond to this ratio is overestimate compared to that obtained by model experiment. It is important to consider extention of breadth and draught ratio in the weather criterion to accomodate ships with ratio larger than 3.50. The effective wave slope coefficient obtained by formula of weather criterion seems to be overestimate when applied to a ship with large breadth and draught ratio. It is necessary an advance investigation of alternative method to determined the effective wave slope coefficient for ships with breadth and draught ratio larger than 3.50. The safety level recommended by IMO in draft of second generation intact stability criteria is more conservative compared to the present result. This means that the minimum safety level can be used for the present subject ship as the minimum safety level.

5. REFERENCES

1. Paroka, D., 'Geometry Charateristics and its Impact on Stability of Indonesian Ro-Ro Ferriees', Jurnal Ilmu Pengetahuan and Teknologi Kelautan, Vol. 15 No.1, February 2018.

2. IMO, 'International Code on Intact Stability Resolution MSC 267 (85), London, 2008.

3. Deakin, B., 'Evaluation of the Roll Prediction Method in The Weather Criterion', International Journal of Maritime Engineering. Vol. 150 No. 2, 2008.

4. Paroka, D., Asri, R., Rosmani, and Hamzah, 'Alternative Assessment of Weather Criterion for Ships with Large Breadth and Draught Ratios by Model Experiment: A Case Study on an Indonesian Ro-Ro Ferry', International Journal of Maritime Engineering (accepted for publication).

5. Francescutto, A., Serra, A., and Scarpa, S., 'A Critical Analysis of Weather Criterion for Intact Stability of Large Passenger Vessels', Proceedings of 20th International Conference Ocean, Offshore and Arctic Engineering, Rio de Janeiro, Brazil 3 – 8 June 2001.

6. Ishida, S., Taguchi, H., and Sawada, H., 'Contemporary Ideas on Ship Stability and Capsizing in Waves: Evaluation of the Weather Criterion by Experiment and its Effect to the Design of a Ro-Pax Ferry', Springer, London, 2011.

7. Borisov, R., and Luzyanin, A., 'An Approach to Assess the Excessive Acceleration Based on Defining Roll Amplitude by Weather Criterion Formula with Modified Applicability Range', Proceedings of the 12th International Conference on the Stability of Ships and Ocean Vehicles, Glasgow, UK, 14 – 19 June 2015.

8. IMO, 'Interim Guidelines for Alternative Assessment of Weather Criterion', Document MSC.1/Circ.1200, London, UK24 May 2006.

9. IMO, 'Finalization of Second Generation Intact Stability Criteria : Information Collected by the Correspondence Group on Intact Stability Regarding Second Generation Intact Stability Criteria', Document SDC 3/INF.10 Submitted by Japan, London, 13 November 2015.

10. Umeda, N., Sakai, M., Maki, A., and Matsuda, A., 'Nonlinearity in the Effective Wave Slope Coefficient for a Low Freeboard Ship', Proceedings of the 28th Conference of Japan Society of Naval Architects and Ocean Engineers, 2019.

11. Rudakovic, S., and Backalov, I., 'On Application of Standard Methods for Roll Damping Prediction to Inland Vessels', Proceedings of the 16th International Ship Stability Workshop, Belgrade, Serbia, 5 – 7 June 2017.

12. IMO, 'Development of Second Generation Intact Stability Criteria: Vulnerability Assessment for Dead-ship Stability Failure Mode', Document SDC1/INF.6 Submitted by Italy and Japan, London, 15 November 2013.

13. Paroka, D., Muhammad, A.H., Rahman, S., and Azis, M.A., 'Operational Limitation of Indonesian Traditional Wooden Boats in the Framework of the Second Generation Intact

Stability Criteria', Proceedings of SIDI International Conference, Surabaya, 2 – 3 September 2019.

14. Sato, Y., Taguchi, H., Ueno, M., and Sawada, H., 'An Experimental Study of Effective Wave Slope Coefficient for Two Dimensional Model', Proceedings of the 6th Osaka Colloquium on Seakeeping and Stability of Ships, Osaka, Japan. 26 – 29 March 2008.

6. AUTHORS BIOGRAPHY

Daeng Paroka is a lecturer and a researcher at Department of Ocean Engineering Hasanuddin University Indonesia. His subject is stability of ships and others floating structures. He has conducted research regarding probability assessment of ship stability in beam seas, developed an alternative assessment of weather criterion applied to an Indonesian Ro-Ro Ferry. He also investigated effect of righting arm characteristics on roll motion on beam seas to determine minimum requirement in stability point of view for ships with large breadth and draught ratios.

Andi Haris Muhammad is a lecturer and a researcher at Department of Marine System Engineering Hasanuddin University. His research topic are ship resistance and ship maneuvering. Recently he has joint in the research about stability of Indonesia ro-ro ferries.

Sabaruddin Rahman is a lecturer and researcher at Department of Ocean Engineering Hasanuddin University Indonesia. Main topic of his research mainly relates to ocean wave modelling and prediction correspond to safety of ship operation.

Rosmani is lecturer and researcher at Department of Naval Architecture Hasanuddin University Indonesia. Her subject is ship resistance and propulsion. She has taken part in the research of aternative assesment of weather criterion for Indonesian ro-ro ferries.

Muhammad Akbar Azis is a magister student at Department of Naval Architecture Hasanuddin University Indonesia. The topic of his magister research is stability evaluation of Indonesian traditional wooden boats using the second generation intact stability criteria.

Ardedi Yusuf is a bachelor student at Department of Naval Architecture Hasanuddin University Indonesia. His bachelor final research correspond to implementation of the weather criterion to an Indonesian ro-ro ferry supported by model experiment.

STABILITY ASSESSMENT OF HATCHCOVERLESS RIVER-SEA CARGO SHIP

S Anggara, M A Kurniawan, S Rosihan and **D R Mauliani,** Research and Development Division - Biro Klasifikasi Indonesia, Indonesia

SUMMARY

Hatchcoverless General Cargo ships is designed to carry exposed cargo while operating due to certain reason. IMO was disseminated the Interim Guidelines for Open-Top Containership by IMO MSC/Circular. 608/rev.1 provided a set of requirement for the design of Open-top Containership. In these Interim guidelines one of the requirement set is related to the stability of the ship in all conditions, intact and damage. Furthermore, in general, the application of these interim guidelines is intended for this type of ship that operate in unrestricted service navigation. The Hatchcoverless General Cargo ship have the same typical design with Open-top Containership, while one or more of the cargo holds need not be fitted with hatch covers. Since, the intended application is generally for unrestricted service navigation, it becomes questionable if applied in restricted service navigation, e.g. river sea service navigation. This work will assess the Hatchcoverless General Cargo ship that has river sea service navigation. This study is working on stability analysis of this chosen type of ship. A sample model was tested by numerical method in intact and damage condition. Which means that computer simulation is adopted to represent dynamic responses of one system by the behavior of another system modeled after it. Dynamic roll response was captured to check allowable downflooding point. The result confirmed and compared with the references regulation. The results found that the model fulfilled criteria in intact condition, but for flooded cargo hold condition showed an inappropriate result. Furthermore, assessment for this type of ship should be reviews against the damage case.

Keywords *: Stability, Hacthcoverless, River- Sea*

NOMENCLATURE

B	Moulded breadth (m)
D	Moulded depth (m)
LOA	Length overall (m)
GM	Metacenter height (m)
GZ	Righting lever (m)
Hs	Significant wave height (m)
T	Draft (m)
Tp	Peak period (s)
Tz	Zero upcrossing period (s)

1. INTRODUCTION

The history of means of maritime transport are evolved as technology development. As well for the type and shape of ship, today it is introduced another change, namely the advent of the open top cargo ship. This type of ship is designed to have large opening without hatchcover above its cargo hold. Thus, the cargo space within the ship is open to the elements. The typical design was actually applied a long time ago in ancient age for carrying stock in Mediterranean. Now it is used again with same concept to catch the same advantage. The main advantage claimed for this design are:

- The considerable weight of hatch covers has been eliminated, thus increasing the deadweight.
- Since the hatch covers were located high in the ship, their removal (removal of their weight) significantly improves stability.
- The elimination of the hatch covers also excludes the need to open and close same. This speeds up port turn-around time and has the potential for reducing cargo operation costs.
- The maintenance of hatch covers, hatch clamps, coaming gaskets, and other hatch securing gear becomes unnecessary.

Disadvantage of this type of design may coming from high risk while on duty. The cargo may not safe and capable to damage by weather, its main reason for the ships only and commonly used for carrying bulk cargo which is not significantly affected by water. The worst risk is that ship may have serious issue in stability, especially when water enter into cargo hold and induce mass shifting and collapse. The risk probably increase when ship operated in oceangoing where environmental condition become harsher and rainfall become higher.

2. REGULATION OVERVIEW

Explicitly ILLC (International Load Line Convention) does not cover, even does not allow it, but implicitly it offer exemption for unusual design. Chance is opened for every flag administration to considered any novel design with emphasize in depth research and serious study to minimize the possible risk. Extensive model test in laboratory shall be carried out to assess ship behavior in seaway, particularly for green water phenomena. In addition, the administration which allow the hatchcoverless design shall communicate it to IMO (International Maritime Organization) if the ship has international voyage.

IMO Maritime Safety Committee (MSC), at its sixty-second session, May 1993, approved interim guidelines for open top container ship which were disseminated by MSC/Circ.608 [1]. This guideline was later adopted in 1994 and still under development up to now and should be revised in the future on basis of relevant proposal from MSC committee member. The circular provides set of requirements for the ship type as may resume as follow:

- Procedure of model test on concern of green water phenomena
- Bilge dewatering system
- Intact and Damage Stability including minimum freeboard
- General and local strength in flooded condition
- Fire protection requirement

Initiated from container-ship, those requirement should be applicable for hatchcoverless cargo ship which has similar concept of design. Thus, Det Norsk Veritas (DNV) has adopted and applied the circular not only for container ship but also General and Multi-purpose dry cargo ship. This regulation is prevailed when at least one hatch cover completely or partially removed during voyage. The ship which is undergone this requirement shall meet the provisions of IS Code 2008 General Criteria for intact stability. The additional requirement has been included that ship shall be able to survive in condition of cargo filled by water in intact condition, namely flooding intact, with a level of water surface 2 m above tank top [2].

Global strength shall be also assessed against still and wave bending moment and shear force, particularly in amidships area. The scantling shall be reduced by corrosion effect which depend on area where structure placed. The hull response due to torsional moment have to be taken into concern since the rigidity of hull is reduced with deck opened.

In term of inspection, survey by administration or Recognized Organization (RO) shall spend more attention to dewatering system and its redundancy. Meanwhile, the structural and hull survey should reflect that they are exposed to the sea atmosphere.

All regulation above are intended for oceangoing operation. It may not suitable for ship operated in river-sea going with more benign environment condition. Some requirement may be adjusted to obtain optimum value. Some independent standard is proposed by administration in basis of their local environmental condition. For example Regulation of river sea vessel belong to Directorate General of Shipping, Government of India [3]. It regulates design and operability aspect, i.e. from construction, stability and survey. River-sea Vessel in this application means that all type of merchant ship, including but not limited for hatch coverless cargo ship, which does not operate beyond territorial water of this states. Specifically the operating area is divided into 4 category that later namely as Type 1, 2, 3 and 4. They may be defined as type 1 and 2, when ship operated in *shore to shore operation* and *nearby port service* respectively, and as type 2 and 3 when ship operated in *restricted coastal* and *unrestricted coastal service* respectively. The category which being used as appropriate reference for this study may come from category 1 and 2. For type 1 and 2, Load line and freeboard of hatchcoverless ship shall account the height of hatch coaming from waterline based on formula in reference [3]. Meanwhile for stability shall comply to following stability criteria.

- C1: GM shall not be less than 0.3 m in any loading condition
- C2: The area under GZ curve up to the angle at which waterline reach hatch coaming shall not be less than 0.055 m-rads.

Secondly, the structural strength shall not be below that require for a sea environment where the maximum significant wave height is 2 m. Wave bending moment reduction factor is introduced to proportionally scale load from ocean going seaway to restricted operation area. Reduction factor (R_{wbm}) are 0.65 and 0.70 for type 1 and type 2 respectively. The hull girder section modulus may also be reduced maximum 0.775 and 0.800 respectively as well. Allowable stress shall not be exceed 175 N/mm^2 for normal strength material.

Refer to rules inland waterway [4] which is applicable for all type of ship operated inland waterway, loads is derived from navigation range coefficient, where the range are classified to still / smooth water, strong current / certain roughness surface (Hs 0.6m), semi maritime/ lake (Hs 1.2m to 2.0m). Certain wind speed to calculate heeling moment will also depend on this range. Wave bending calculation shall also be taken from navigation coefficient n (0.8xHs). Requirement of stability level is specifically being set up based on the type of ship. For example, only maximum GZ and area under GZ are required for oil tanker. Additional damage requirement is applied only for container ship with length of 100m and above. Special for pontoon, only minimum residual freeboard required for at least 0.3 m where 10 degrees heeling does not lead deck immerged. As for bulk carrier, attention to be given when ship heel to more than angle of repose of its cargo.

China classification Society (CCS) in reference [5], set a requirement for minimum freeboard of open top container ship operated in restricted area. Either by seakeeping model test if scupper provided or by geometrical calculation. The model test is simulated by Hs of 3.5 m (category 3) and 5.5 m (category 2) for navigation range of sheltered area and up to 20 nautical miles respectively if scupper or freeing port. As for geometrical calculation, prescriptive based formula as faction of ship length is used. Furthermore, flooding intact is also required as cargo hold shall be filled with water up to70 % and up to 55% for category 2 and 3 respectively.

Compared to river-sea going, the requirement of ocean going is developed by basis of more harsh environmental condition, which Hs of approximately 8.5 m at the most unfavorable realistic Tz. The Pierson-Moskovitz, JONSWAP, or Bretschneider wave spectrum may be used for this circumstance. However, DNV provide exemption for ship operating in restricted area only. Other spectrum and Hs may be proposed under DNV approval. This similar to river-sea vessel rules which is applied much lower Hs (2 m) to represent average condition of sea-river in India. The criteria stability of IS Code majorly to be vanished in River-sea stability criteria, However GM criteria is increased for compensating other criteria excluded. Damage condition is also excluded from this local criteria and flooding intact is not considered.

2. METHOD

This study is focused on applicability of regulation to typical design hatchcoverless ship operated in Indonesia river-sea. The IMO interim Guideline MSC/Circ.608, DNV rules and RSV rules of India Government are tested to model in term of stability aspect. Seakeeping performance respect to Indonesia sea-river environmental condition is also analyzed to support the result.

2.1 MODEL

Model was adopted from actual cargo ship carrying coal in bulk which was operated in Mahakam river, Indonesia. Main dimension is as follow:

LOA	:	68.50	m
B	:	13.00	m
D	:	4.20	m
T	:	3.12	m
Disp	:	2480.9	ton

Visualization can be seen at Figure 1. All arrangement has been modeled as well e.g. superstructure and hatch coaming respect to general arrangement in order to capture the wind effect. The upper surface of hatch is considered as flooding point. All tanks was modeled, while cargo hold assumed as compartments when damage stability simulated.

This model has been verify to actual ship with certain correction based on reference [6] that the deviation of displacement is not allowed more than 5 % and 1 % of parameter related to mass distribution. In addition, Cross curve gap is to be not more than 5 % in all degree inclination.

Figure 1. Hatchcoverless ship model

2.2 STABILITY

Each condition of Ballast, intermediate (cargo hold filled to 25%, 50%, 75%) and full cargo were investigated for departure and arrival cases. Wind heeling moment is determined by considering 16 knot of wind speed. Stability in intact condition was checked against international and local code as per references.

Damage condition to meet requirement MSC.608 was also analyzed. Both damage probability as per SOLAS Chapter. II-1 Part. B explained in MSC res. 281 (85) and deterministic are investigated [7]. Basically, this damage procedure is used to check subdivision requirement especially for passenger ship with more than 12 passenger, thus its applicability to hatchcoverless is questionable and will be answered in this study. Damage requirement has been into forced since 1929 using deterministic method, Start from 1978, amended on 1981 and 1983, probability based method is adopted through IMO res.265 (VIII). This regulation is then expanded to cargo ship with subdivision length above 100 m.

Specifically in deterministic, outward cargo to be assumed leak alternately from fore to afterward of the ship, then continued with 2nd level damage which two side by side tank are leak at once time. Thus total of 156 scenarios to be run.

2.3 SEAKEEPING

Based on [8], sea river navigation in Indonesia is presented by Hs 1.25 m with Tp 5.87 s, equivalent to Beaufort 4 with wind speed range to 11 to 16 knots. But, in order have comprehensive result, sea state is variated from Hs 0.2 m to 3 m with adapted Tp, see Table 1. Those data was utilized to capture roll response in seakeeping analysis. The result then compared to maximum heel angle allowed by regulation and become benchmark to describe local river-sea condition compared to oceangoing seaway.

In order to model environmental condition, the standard wave spectrum Pierson Moskowitz was used. Mass distribution become very important to determine roll radius gyration. This parameter may be compatible to GZ in static stability to gain restoring force. Response amplitude operator (RAO) was then multiplied to wave spectrum to obtain ship motion response. Strip theory has been used with 180 degree of wave heading and 30 degree spacing.

Table 1. Sea state variation

Hs	Tp	Hs	Tp
0.25	2.629	1.75	6.947
0.5	3.717	2	7.425
0.75	4.552	2.25	7.874
1	5.256	2.5	8.305
1.25	5.878	2.75	8.7
1.5	6.438	3	9.091

3. DISCUSSION

Initial investigation is to check model to IS Code General Criteria. In some cases, i.e. in full load condition with maximum draft, the model is harder to comply each criteria compared to smaller draft since the parameter such GM, area under GZ become smaller in higher draft. In other hand, it induce smaller freeboard where coaming is closer to waterline. It restricts allowable heeling angle of the model. Figure 2 illustrates IS codes compliance of the model. It shows the ratio of actual value to requirement of 6 (show in axis) criteria in IS Code part. A. Ch.2.2 [9]. There are 12 load cases tested for each criteria that is sorted from higher to smaller draft. Smaller draft leads to smaller ratio of compliance and negative histogram means that model does not comply. The model does not able to comply criteria number 3 which require area under GZ between heel angle 30 and 40 degree not less than 0.03 m-rad especially for higher draft. This happen because waterline reach upper coaming with heel angle less than 30 degree (20-25 degree). Criteria 3 would be passed if only draft set not more than 2.4 m, means ship need to discard 950 tons of cargo. It will significantly cut off the economical value of the ship and it is not good option for ship owner. Furthermore, minimum freeboard of 0.7 m is calculated based on ILLC, thus the similar draft, i.e. 2.42 m has match to maximum value required for IS Code criteria.

The second assessment is made against local criteria of river-sea vessel. All criteria is satisfied by all 12 load cases. The criteria is also checked through draft increment and it is found that ship fail to first criteria C1 at draft 3.7 m. Instead of reducing cargo respect to the MSC 608, the draft may be increased up to 0.68 m or equivalent to 510 ton of cargo weight. The same draft is resulted by prescriptive formula of minimum draft from the same regulation. It gives understanding that the freeboard formula is derived from those stability criteria. However, the maximum heel is to be allowed only up to 15 degree instead of 30 degree. See Table 2.

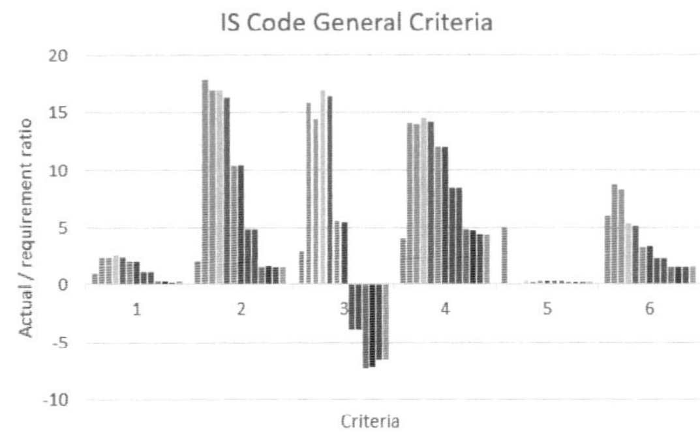

Figure 2. IS Code compliance

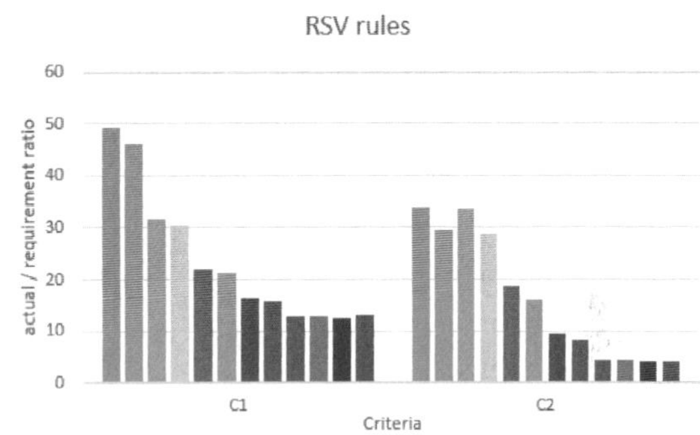

Figure 3. RSV rules compliance

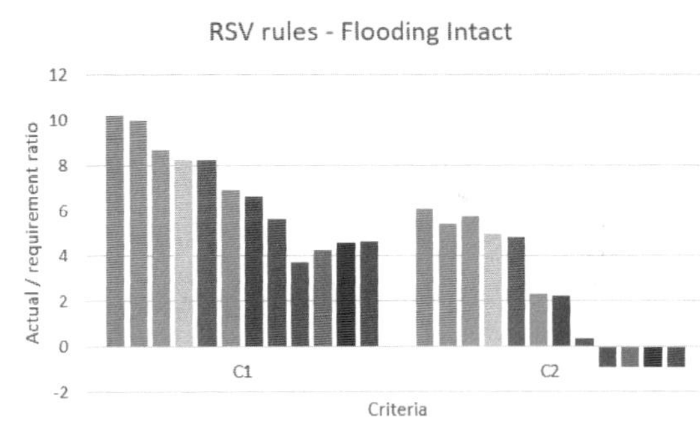

Figure 4. RSV rules compliance in flooding intact condition

Hereafter, The applicability of flooding cargo hold in intact condition as per DNV requirement is tested to the model. Cargo hold is filled water up to 2 m above tank top to check survivability of ship in case of bilge discharge system failed. The model clearly will not capable to comply IS Code since in normal condition it does not.

Table 2. Criteria and freeboard relationship

Draft [m]	3.1	3.2	3.4	3.6	3.8	4	4.2
Displacement [ton]	2481.0	2627.0	2734.0	2907.0	3081.0	3254.0	3427.0
Downflooding Angle (deg)	21.1	19.2	18.3	16.1	14	11.7	8.8
Coaming Height [m]	2.3	2.2	2	1.8	1.6	1.4	1.2
Initial GM > 0.3 [m]	3.83	3.72	3.564	3.413	3.072	2.99	2.97
Area under GZ curve > 0.05	0.242	0.22	0.158	0.108	0.045	0.028	0.006
Status	ok	ok	ok	ok	not ok	not ok	not ok

Instead of using IS Code, RSV rules is then applied. Figure 4 shows that in higher draft, area under GZ curve (C2) become smaller. 4 load cases with highest draft, that in normal condition satisfy the criteria, fail when flooding intact procedure applied in existing design. As maximum draft allowed by river-sea rules is 3.7 m in normal case, in flooding intact that draft is obtained by load case departure with 75% cargo, which means in flooding intact the model shall discard 25% of cargo content to comply RSV rules criteria. Then, 2 m water level may be too high if applied for RSV rules, since it may concern for ship with length above 100 m.

From full load draft of 3.11 m to 3.7 m there is 510 ton remain weight can be added, 510 ton water equivalent to 1.01 m or 20% of cargo hold depth. The value may become maximum water level and more applicable if provided in percentage of cargo hold depth. However different ship should have different percentage but does not deviate significantly. CCS has recommended water level 55% of cargo hold depth, but it specifically for only container ship.

Damage condition has been investigated by probabilistic damage to check subdivision requirement to the model. The result then supported with deterministic procedure which each outward tank damaged alternately. The result show that attained subdivision index A is less than required subdivision index R which means the model has high risk to collapse when a compartment, i.e. cargo hold is leaked.

Required Subdivision Index (R)

Cargo ship

Ls >100m $\quad R = 1-(128/(Ls+152))$

$R = 0.448276$

LS<=80m $\quad R=1-(1/(1+(Ls/100)X(R0/(1-R0)))$

$R= 0.393939$

Attained Subdivision Index (A)

$A = 0{,}4\ As + 0{,}4\ Ap + 0{,}2\ Al$

$A = 0.206353$

As = 0.206352835

Ap = 0.206352835

Al = 0.206352835

As, Ap, and A1 represent deepest, partial and intermediate condition of load line. The Ls of the model is 45.71 m thus R = 0.39, greater than A = 0.206.

Deterministic by 156 scenarios show the same result, when cargo hold damaged ship in full load condition will experience capsize since its displacement is not sufficient to float. However in intermediate and ballast condition, ship remain survive. Then when 2nd level damage tested, the intermediate case no longer sustain but the ballast case still. This explain that even only a cargo hold damaged, the ship will capsize if having full cargo. The solution is to limit size of cargo hold.

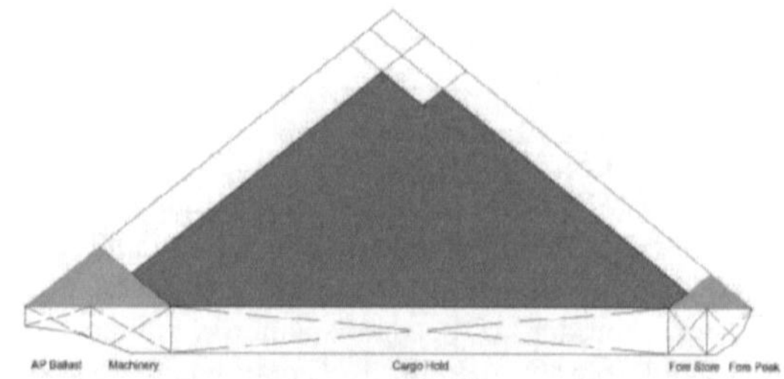

Full Load

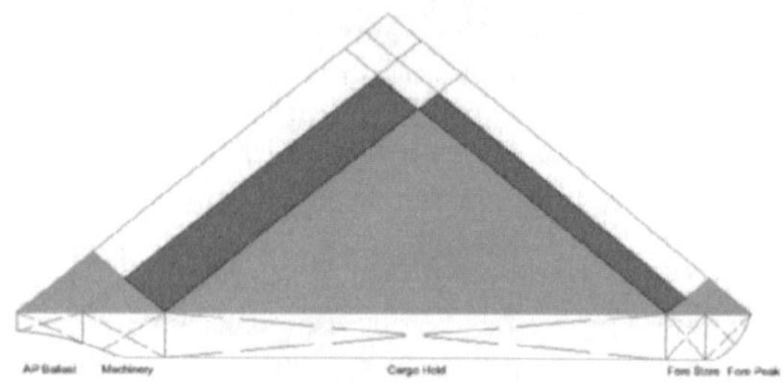

Intermediate Load

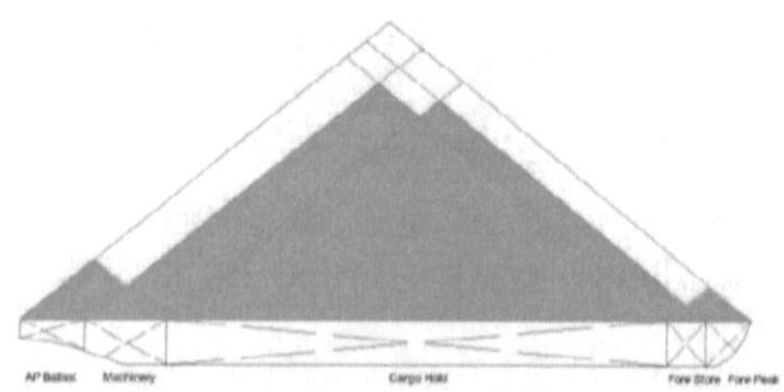

Ballast Load

As for seakeeping, roll response is checked within variated sea state. MSC 608 recommend to use Hs of 8.5 m in seakeeping analysis. In that condition, the model gives response of 33 degree roll angle in wave heading 90 degree. It is match with assumption of 3rd criteria IS code that requires certain area under GZ curve between 30-40 degree heeling, which means that flooding point shall in such position that results minimum heel angle of 30 degree to reach the position.

Furthermore, as discussed above, minimum freeboard by RSV allow model to have up to 3.7 m draft. In this condition upper coaming will be reached in only 15 degree heel angle. In that draft and that heel angle, minimum of Hs 0.75 m degree are needed. Thus it may be a technical background behind RSV rules development for load line and freeboard.

In normal draft of 3.1 m, downflooding angle is about 21 degree. Seakeeping analysis show that the model experiences that rolling angle if 1 to 1.25 m of Hs is applied. This is significant wave height of river sea waterway of Indonesia. Meanwhile 2 m Hs results roll response of 23.5 degree.

As per CCS requirement, minimum freeboard can be determined by knowing maximum roll response in seakeeping analysis. CCS recommend Hs of 3.5. In this case model gives response of about 25 degree, which exceed to downflooding angle. When draft reduced to 2.9 m then downflooding point cannot be reach by Hs 3.5. If using prescriptive formula, freeboard minimum is 1.85 m (0.0275Lwl) or draft become 3.55, smaller draft than RSV rules.

4. CONCLUSION

After comparing the regulation above, it may be understood that MSC 608 should not applicable for hatchcoverless cargo ship operated in river-sea. Since the background behind the development is specifically for oceangoing container ship. Then DNV has tried to adopt the regulation for general cargo ship but still apply condition while ship operating in open sea. The special statement is provided to cover case if ship operated in more benign environmental area, in this circumstance designer is allowed to adjust sea state to be more suitable to apply. However it is found too hard if the stability criteria is remain according to IS code. Discharging some cargo can increase stability but it is not a good option for ship owner. In other hand, damage case procedure required by SOLAS Ch.II-1 Pt. B is known too hard for typical hatch coverless with very wide and long cargo space. The addition of transverse bulkhead may not help significantly since water can be transport through upper coaming.

Meanwhile for typical river sea vessel which has small freeboard and limitation to roll, RSV criteria should be more suitable. The model meet well with the criteria. Even, increasing the draft is remain accepted respect to minimum freeboard requirement. In addition, flooding intact application is possible to be applied for small freeboard with several adjustment especially for the assumed water volume inside the hold.

Seakeeping analysis may become alternative to determine minimum freeboard by calculating ship maximum rolling response. The angle when upper coaming contact to water is considered as minimum freeboard. RSV is recognized using 0.75 m of Hs for determining allowable draft. However when considering Indonesia environmental condition, Hs of 1.25 m, higher required freeboard is needed.

Writer have recommendation for advanced research to add number of sample model to obtain more comprehensive and representative result.

5. REFERENCE

[1] IMO (International Maritime Organization), 'Interim Guidelines for Open-top Containerships', *MSC/Circular 608/Rev.1*, 1994

[2] DNV-GL, (Det Norske Veritas – Germanischer Lloyd) 'Ships Designed without Hatchcover Hatchcoverless (Part 6 Chapter 5)', *Rules for Classification*, 2017

[3] Ministry of Shipping of Government of India, 'River-Sea Vessel (RSV) Notification', *Rules for Construction*, 2013

[4] BKI (Biro Klasifikasi Indonesia), 'Design Load Principle', *Rules for Inland Waterway Vessel*, 2015

[5] CCS (China Classification Society), 'Additional Requirements for Open-Top Container Ships', *Rules for Classification of Sea-Going Steel Ships*, 2018

[6] BKI (Biro Klasifikasi Indonesia), 'Pedoman untuk Sertifikasi Sistem Komputer Pemuatan', *Pedoman untuk Klasifikasi dan Konstruksi*, 2015

[7] IMO (Iternational Maritime Organization), 'Explanatory Notes to the SOLAS Chapter II-1 Subdivision and Damage Stability Regulations', *MSC.281 (85)*, 2008

[8] Kurniawan M.A., Preasetyo F.A., Komariah S., 'Indonesian Seastate Condition and Its Wave Scatter Map', *MASTIC (Maritime Safety International Conference)*, 2018

[9] IMO (International Maritime Organization), 'International Code on Intact Stability'. 2008

STUDY ON THE ALTERNATIVE METHOD OF NON-DESTRUCTIVE TESTING FOR FIBER REINFORCED POLYMER

V N Garjati, H R Ardyanto, M Aulia, Research and Development Division - Biro Klasifikasi Indonesia, Indonesia

SUMMARY

The usage of Fiber Reinforced Polymer (FRP) as a main construction material in ship's hull is increasing every year. Due to the combination of its lightweight characteristic and excellent mechanical properties, FRP materials can be used in speed boat, high speed craft, passenger ship, and any other fast ships. The excellent mechanical properties can be achieved during the shipbuilding process. However, on the recent years, there are many accidents happened on FRP Ships. In fact, many FRP ships are built without the supervision of Classification Society, so that its material properties and quality cannot be ensured. It is not easy to assess the quality of installed materials on an existing ship, since it is not feasible to conduct a destructive test to determine mechanical properties; on the other hand, it is very important task. Hence, non-destructive test method is more preferable to determine quality of materials, considering its practicability for portable examination, and only small damage caused by this method. Modal Testing (Vibration Test) is one of the promising non-destructive test method to assess material properties. Furthermore, the material properties resulted from Modal Testing were compared with Tensile Testing as a conventional testing method. The comparisons then being analyzed and presented in this paper. Recommendations for further works were also included.

***Keywords** : fiber reinforced polymer, modal testing, vibration testing*

1. INTRODUCTION

Composite is a combination of two or more types of materials in order to achieve the desired properties of material. One of the most common composite used in the world is fiber reinforced polymer (FRP). In FRP, glass fibers are attached to polymers material, the combination resulting a material which possess a high strength but a lower density compared to steel. For the past years, it has been used as a construction material in ship's hull. Therefore, because of its lightweight characteristic, FRP is commonly used as a hull construction material for various small and high-speed crafts.

To ensure that the FRP material achieve the excellent mechanical properties on its application in a ship's hull, the supervision from Classification Society during shipbuilding process is important. Lack of materials quality may bring some drawbacks for the voyage, as well as ship's safety. Moreover, they may carry many passengers. The National Transportation Safety Committee (KNKT), an Indonesian government agency charged with the investigation of air, land, rail, and marine transportation safety deficiencies, published that some incidents and accidents on FRP ships are already happened in Indonesia.

An accident happened on MV. Dumai Express 10 on its voyage to Pulau Iyu Kecil, Tanjung Balai Karimun, Riau Island [1]. MV. Dumai Express 10's hull is constructed by FRP materials, and the accident is happened after 10 years of shipbuilding process. The shipbuilding process was not supervised by any Classification Society. The process was done, and supervised by its own shipbuilder. Moreover, all of materials used in the process did not have the approval from any Classification Society. In fact, both of materials used and the building process play an important role in ensuring the strength of construction. During its voyage, the big wave hit the side wall plate of the construction and the water gets into the ship through the damaged wall. After some investigations, it is known that the damaged wall materials possess a brittle characteristic due to the imbalance composition of resin and catalyst. The accident caused MV. Dumai Express 10 sank, 42 passengers died, and another 33 passengers were not yet found.

In order to ensure the safety during its voyage and prevent the damage of construction, a ship shall be inspected by Classification Society. The mechanical properties on the construction which represent the condition may be verified. The conventional tensile testing may not be used for examining the existing ships since it may damage some parts of ships construction and needs more time as well. On the existing ships, the unavailability of material data sheets is also becoming an issue in determining and predicting the mechanical properties of FRP materials.

Non-destructive test (NDT) is one of the method that can be used to verify the mechanical properties of material with a very minimum damage left on the constructions. Since it doesn't need any specific shape and dimensions of samples, it can be done in a very short time and process. Considering its practicability for portable examination, NDT become a promising method to assess material properties on ships. It is not easy to assess the quality of installed FRP materials in ships using NDT method, on the other hand, it is a very important task.

2. REQUIREMENTS FOR FRP SHIPS

BKI classification admission procedure for new building FRP ships is as follow:

1. Application forms is submitted to BKI
2. Approval for document and drawing
3. **Approval for material and component**
4. Verification for ship's construction and testing
5. Shipboard trial
6. Final survey
7. Survey report and class certificate is released

Approval for material and component through material and component certification is done by material testing and strength test in accordance with BKI Rules for Fibre Reinforced Plastics Ships (Pt. 3, Vol. III). These tests used for hull construction of FRP ships using FRP laminates and sandwich laminates. The material tests for laminates include thickness of moulding, barcol hardness, glass content ratio, bending strength, modulus of bending elasticity, tensile strength, and modulus of tensile elasticity. The requirements of material mechanical properties based on glass content are in accordance with BKI Rules for Small Vessel up to 24 m (Pt. 3, Vol. VII).

On the other hand, the method for material assessment in classification admission procedure for existing FRP ships is currently being developed. As mentioned before, the conventional tensile testing may not be used for examining the existing ships since it may damage some parts of ships construction and needs more time as well. On the existing ships, the unavailability of material data sheets is also becoming an issue in determining and predicting the mechanical properties of FRP materials.

3. FRP MATERIALS

Composite is a material which consist of two or more different types of materials in order to achieve the desired combination properties. In general, this material is composed from matrix which has low strength and added by reinforcement which has higher strength in order to achieve excellent mechanical properties. In Polymer Matrix Composite (PMC), polymer is used as matrix with fiber (ceramic) as a reinforced material. There are some types of commonly used fiber, e.g., glass, carbon, aramid. Glass Fiber-Reinforced Polymer (GFRP) or often called fiberglass may also be used as a construction material in ship's hull. The glass fiber has a relative high strength to produce a composite with a high specific strength. This kind of composite also has a high chemical inertness which makes it suitable in any corrosive environment.

In composite material, metals and polymers are commonly used to be matrix because of their toughness. This is related to the functions of matrix. First, as a binder for fibers while distribute the stress to the fibers. Second, it also protects the fibers from the surface damage resulted from mechanical abrasion and chemical reaction. On top of that, matrix prevents the propagation of crack from one fiber to another. Polyesters and epoxies are commonly used as a matrix for fiberglass in shipbuilding.

Fibers orientation, concentration, and distribution have significant influence on the mechanical properties of fiber-reinforced composites. Continuous fibers are generally aligned, whereas discontinuous fibers may be aligned, randomly oriented, or partially oriented. Mechanical responses to this type of composite depend of several factors to include the stress-strain behaviours of fiber and matrix phases, the phase volume fractions, and, in addition, the direction in which the stress or load is applied [2].

Composites which have the aligned fibers are highly anisotropic and dependent on the direction in which they are measured [2]. Strength on the direction which is parallel to the fibers is the maximum strength, while the perpendicular direction to the alignment is the minimum strength of the composites. On the other hand, the randomly oriented fiber composites are highly isotropic.

In structural application, laminate composite is commonly used. Laminate consists of aligned composites (i.e., fiberglass). This aligned composite is stacked each other in any directions which can be modified, therefore it is very possible to produce the isotropic properties on this material. Fiberglass as a reinforced material can be made from various types of glass, i.e., E-glass and C-glass. Form of E-glass which commonly used as a construction material for ship's hull are woven cloth, chopped strand mat, and woven roving.

Non-metallic materials have different elastic behavior to the metallic materials, since its elastic portion of the stress-strain curve is not linear, as shown as in Figure 1. Tangent or secant modulus is normally used determine a modulus of elasticity.

Metals and ceramics have a constant Young Modulus near the room temperature, while polymers have time and temperature dependent modulus. Because of its different characteristic with metallic materials, comprehensive experimental data are necessary in order to understand the behavior of polymer materials. The stiffness of polymer will vary with time, stress, and temperature; while its toughness is influenced by the design and size of the component, design of the mold, processing conditions and the temperature of use [3]. A slight change of those parameter may affect its basic physical properties.

Strain at yield is an indication of the degree of strain which materials can accept without yielding in typical room temperature [3]. For plastic polymer, yield point is taken as a maximum on the curve, occurs just beyond the linear-elastic region [2]. Stress at this maximum point is the yield strength.

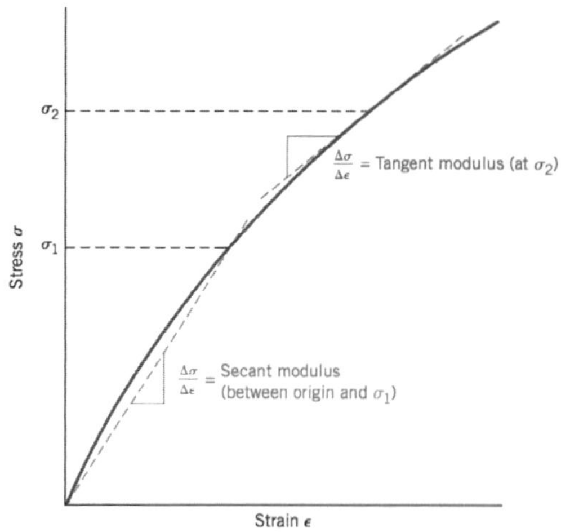

Figure 1. Modulus of Elasticity Determination for Non-Metallic Materials [1]

Materials which has elastic solid criteria has a definite shape and deforms under external forces. It also stores all energy that is obtained from external forces, which will restore it back to the original shape once the force is removed. Materials in viscous liquid state has no definite shape, and also flows irreversibly under the action of external forces. Material polymer response which combines both liquid and solid is termed viscoelasticity. It can exhibit all intermediate range of properties between an elastic solid and a viscous liquid.

4. TESTING FOR FRP

FRP laminate is commonly used in structural ships material. Its light weight compared to steel contribute to easier ships manoeuvring and movement. However, the big waves during the voyage is becoming the thread for this material for its low toughness characteristic. So, the fulfilment of minimum mechanical properties required of materials is important to prevent the material or structural damage. In general, there are two kinds of material assessment and verification, destructive test and non-destructive test.

a. Destructive Test (DT)

Mechanical test can be used to determine the material response on a structure, construction, or component to a given force. Tensile test becomes the most commonly used destructive testing method. It can determine various mechanical characteristic and properties of materials.

The testing involves the increasing tension stress that is given to the specimen until it is breaking while the transition behaviour and its elongation is recorded and converted into an engineering stress strain graphic. The graphic is widely known as a stress strain curve. Stress strain curve, as described on Figure 2, may be used to analyse the mechanical properties of material, such as ultimate tensile strength, yield strength, proportionality limit, elasticity limit, ductility, modulus of elasticity, toughness, and also material type of fracture.

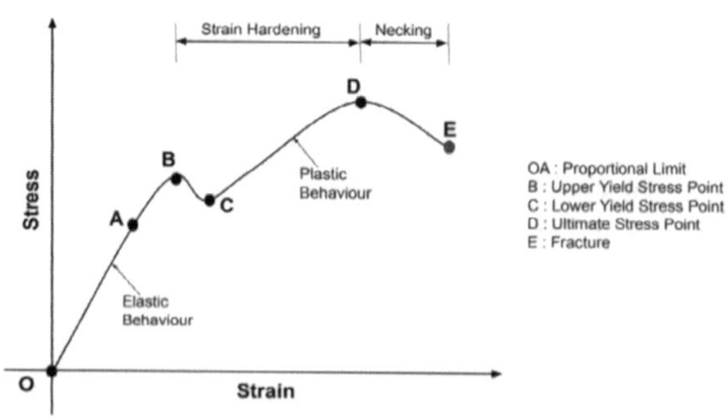

Figure 2. Stress-strain curve of materials

b. Non-Destructive Test (NDT)

On existing ships, verification of material properties shall be done, on the other hand, it will leave some damage on the ships construction when Destructive Test is performed. Therefore, Non-Destructive Test might be needed in the verification. One of the NDT methods that can be applied to laminate composite is vibration testing or modal testing impact method. Modal testing can be used to determine the natural frequency value of a material, which can be related to its mechanical properties.

Vibration is a repetitive motion that can be measured and observed in a structure [4]. Once it is modified, it can be used to define the response of the structure. Vibration analysis is divided into some sub-categories, such as free and forced vibration. In order to obtain the better understanding of structure, free vibration is chosen as a method used since it is the natural response of a structure to some impact [4]. The response of the structure is completely determined by its properties;

therefore, material's mechanical properties plays a big role on its vibration characteristic.

Oscillation is vibration's movement back and forth at a regular speed, while period is a time between every oscillation. Simplest vibration model can be illustrated by mass-spring-damper model. The model consists of simple mass (M), suspended by a known stiffness spring (K), and a damper (C) like a shock absorber, which produces opposing force to the velocity of the mass, as seen on Figure 3.

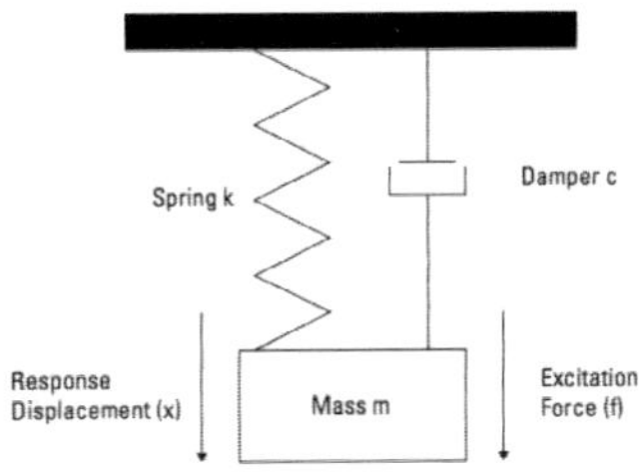

Figure 3. Mass-Spring-Damper Model [4]

Frequency can be calculated by dividing one by the period. For the mass-spring-damper model subject to free vibration, frequency of oscillation determined by the parameters M, k, and C. The oscillation creates frequency as a response to accelerated or impact, called the natural frequency (ω_n). It is measured in cycles per second (Hz).

Instrument for Vibration Testing

Vibration Testing Instrument consists of three main components; they are excitation source, accelerometer, and digital signal analyzer.

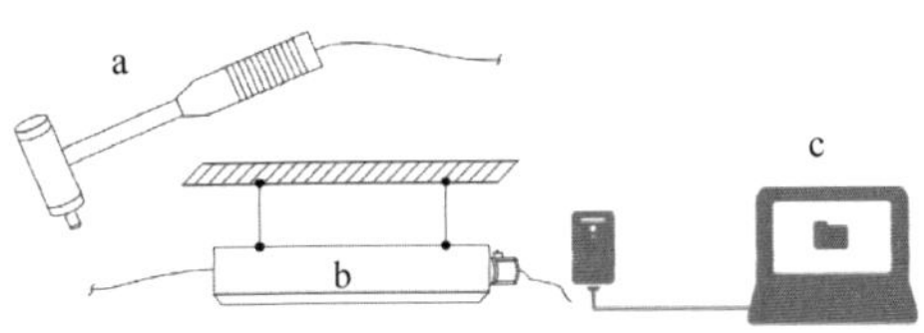

Figure 4. Vibration Testing Instrument Scheme (a) Excitation Source, (b) accelerometer, (c) Digital Signal Analyzer) [5]

Excitation Source

An impact hammer can be used as one of the alternative excitation sources in vibration testing [5]. This hammer produces vibration in a very short time by hitting the structure on some particular points. A sensor, or often called transducer, is attached on its contact tip, so that the force given to the structure can be measured. The given 'mass' can be calculated by measuring the acceleration and the impact force on the hammer's tip and further it will be recorded in the connected PC.

Size of the hammer and hardness of the contact tip control the duration of the impact and thus the frequency. The harder tip the shorter the impact time will be, on the other hand, the softer the tip the longer the impact time.

Accelerometer

Structural material's response to the vibration can be detected by using an electronic sensor. Accelerometer can be used to measure the acceleration response of the structure, or the oscillation, to the excitation given. Then, the responses will be read, processed, and modified into electrical signals. The electrical signal then being analyzed by using time or frequency domain [5].

Accelerometers are chosen based on the range and size of desired frequency. Another important factor for the accelerometers is the distance from the excitation source. It will be affecting the signal detected by the instrument. It is best to keep the accelerometer near the source to make it easier to detect the signal.

Digital Signal Analyzer

Digital signal analyzer on the testing instrument is used to analyze and transform the captured or detected signal by the accelerometer. Signal analysis is generally divided into time and frequency domains; each domain provides a different view and insight into the nature of the vibration [5].

In time domain analysis, the signal is analyzed as a function of time. Response of an impacted structure to the vibration is measured at particular point and plotted into time versus amplitude. Nowadays, vibration measurements are typically plotted in the frequency domain, by using Fast-Fourier Transform, so that the time-domain signals can be processed into frequency domain. It is mainly because of time domain signals are typically complicated and hard to be identified [5].

Figure 5 (b) shows two sine waves of different frequencies summed together. By using frequency domain analysis, it is possible to simplify the multi sine waves as shown in Figure 5 (c). The result of vibration measurement can be plotted into frequency versus acceleration. The record of frequency from impacted structure then be classified into some frequencies. Frequency which has the highest amplitude, is the natural frequency. It can be related and be used to predict the mechanical properties of materials.

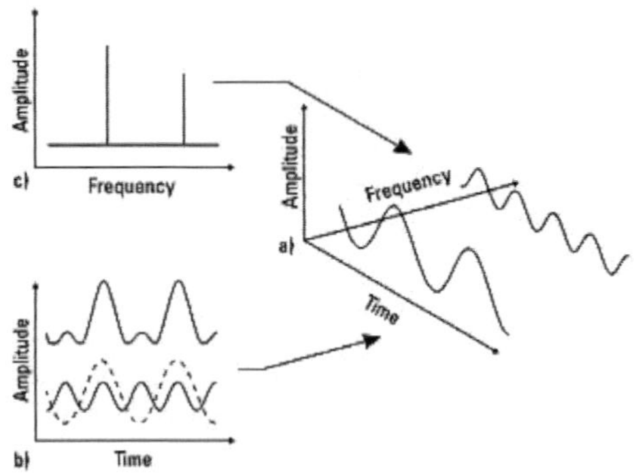

Figure 5. Time and frequency domain

Simple experiment of Young Modulus measurement either in static and dynamic methods, have been performed by Salvatore Ganci [9] by using flexural vibration experiment on thin rectangular bar made from wood. This rectangular bar had been vibrated using sound amplifier until it reaches the desired deviation. This research producing linear curve of frequency and acceleration square. The slope of the curve will produce the value of elastic modulus that can be compared with static tensile testing.

There is also a research which compared Young Modulus from calculation beam theory, static bending test and vibration impact hammer test conducted by Chih Lung Chao [10]. This research using thin wood rectangular shape and producing each different Young Modulus, which will be validated through each consistency ratio of Elastic and Shear modulus. There is still tendency that vibration test result is 1.2% higher than bending test.

In another paper, composite material testing had been defined into two categories which is static and dynamic testing. in order to determine material strength. The experiment was using various material systems such as Cross Plied and Symmetrical Balanced Glass/Epoxy, also Glass/vinylester and Carbon vinylester. Testing method has been arranged for tensile and vibration testing. Then both of the result have been compared and it is found that the value will differ for about $\pm$ 5%.

Those researches are combining several methods to have a better understanding the correlation between static and dynamic testing methods on composite materials.

This work is expected to validate the modal testing as an alternative method of tensile testing substitution on the ships structure, by comparing the testing results for both of static and dynamic testing. Tensile testing and modal testing are used in this work.

5. EXPERIMENTAL METHODS

a. Materials

Materials used as the testing samples in this work:

Type	: fiberglass
Matrix	: Polyester
Reinforcement	: Woven Roving and Chopped Strand Mat
Thickness	: 12 cm
Manufacturer	: PT Arindo Pacific Chemical
Heat Treatment	: T = 60°C, t = 120 minutes

b. Tensile Testing

Tensile testing for 5 samples is performed using Universal Testing Machines Shimadzu EHP, with capacity of 20 tons. Results of the test presented in the form of stress strain curve, and some mechanical properties data such as elongation and yield strength. Further analysis is discussed in Results and Discussion section.

c. Modal Testing

With the modal testing method, the dynamic parameters of the plate will be defined, such as natural frequency, vibration modes, and also structural damping factor. Tools used in this testing methods are as follow:

a. ICP Accelerometer B&K type 4516
b. Dewe Data Analyzer 4T-T-DSA
c. Impact hammer B7K type 8202 with Conditioning Unit B&K type 2626
d. Computer

Modelling for sample geometry is made with DeweFRF program. 5 pieces of composite plate sample (30 x 30 x 1,2 cm) is divided into 36 points of testing. In every point, the excitation is given force with an impact hammer. In testing specimen, the points and its geometry are divided into 36 points, as similar as possible with the modelling geometry.

The accelerometer is attached to one of the points. Before the accelerometer is used for the testing, the validation for the accelerometer and data acquisition system shall be made in order to produce the accurate testing results. The validation for the system was using shaker calibrator Onosokki VK-1100.

This testing method is in free-free state condition, the specimen is on the flexible foam. The data acquisition is already set on maximum frequency of 500 Hz. The excitation was done on each point and the data was recorded automatically.

6. TEST RESULTS & DISCUSSIONS

a. Tensile Testing Results

The results for 5 tensile test specimens is shown in Table 1. Calculation from elongation and tensile strength was made to determine the value of each sample's Young Modulus. These results then were being compared with modal testing to obtain the comparison parameter for these testing methods.

Table 1. Young Modulus from Tensile Testing

Sample	Young Modulus (E)
1	43.2
2	42.9
3	44.6
4	43.2
5	42.9

b. Modal Testing Results

Modal testing was conducted to 5 composite plate specimens. Each specimen was subjected to 36 times of excitation on 36 points and resulted 177 sample data. Each excitation point had been recorded during the test and resulted data in the form of frequency and amplitudes. The first natural frequency on each excitation point is determined by creating a plot of data into the amplitude vs frequency curves [6], as shown in Figure 6. Natural frequency (f_n, Hz) is an effect from an excitation system with unlimited or varied amplitude.

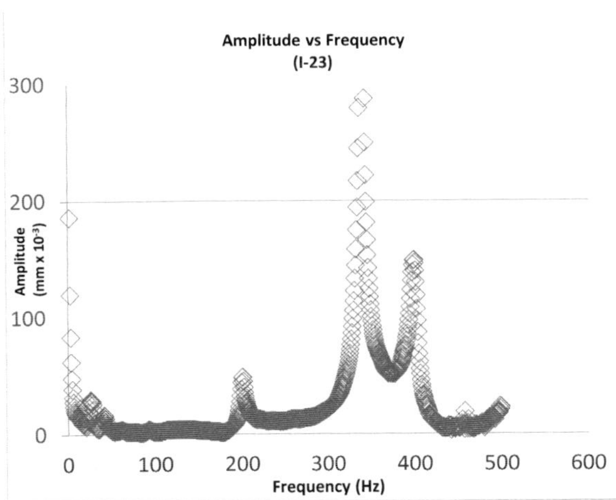

Figure 6. Amplitude vs Frequency Curve on Sample 1, point 23

The numbers of natural frequency on a system is based on the numbers of mode shape created on it. The first natural frequency can be found on the first mode shape since it has the most linear amplitude. It is related to the most suitable condition to determine the maximum stress a structure can hold without any deformations. In a structural design, the maximum permissible stress of material is limited by the numbers of linear elasticity stress.

It is easier to determine the value of first natural frequency on a plotted curve, by determine the peak of the wave from the left is the 1st, 2nd, 3rd natural frequency. In Figure 6, the red mark on the curve is the first natural frequency on point 23 Sample 1. The same method is applied to another 176-sample data. Furthermore, the natural frequency can be used to determine the Young Modulus (E). The Modulus Young value can be defined by using ASTM E1876 [7].

<u>Sample thickness and Young Modulus (E) calculation</u>

The ratio between t/r on the sample is 0.8. Therefore, the sample is categorized as a thin sample. Based on ASTM E1876, Young Modulus can be defined by using Equation 1 [8][11].

$$E = 0.9465\,\rho\,{F_{f,1}}^2\,{L_x}^2\,\frac{{L_x}^2}{{L_z}^2}\,T_1\left(\frac{{L_z}^2}{{L_x}^2}, v\right) \quad \text{... Eq. 1}$$

L_x : dimension in x axis
L_z : dimension in z axis
T : form correction factor

For thin sample category, T values can be calculated by using Equation 2.

$$T_1 = 1 + 6.585\,\frac{{L_z}^2}{{L_x}^2} \quad \text{... Eq. 2}$$

The result for all point and sample is presented on Table 2. Testing specimen's mass and specific gravity is presented on Table 3.

Table 2. Young Modulus from Modal Testing

Sample	Natural Frequency (Hz)	Young Modulus (E)
1	200.14	30.62
2	200.20	30.64
3	200.80	30.82
4	199.07	30.30
5	199.03	30.37

Table 3. Testing specimen mass and specific gravity

Sample	Mass (g)	Specific Gravity (g/m^3)
1	1534.6	14.21
2	1516.6	14.04
3	1551.6	14.37
4	1537.1	14.23
5	1531.2	14.18

c. Comparison of Testing Results

All data from 5 samples resulted from tensile and modal testing is compared. The comparison leads to the approaching of consistent deviation of the two values. By dividing the value of young modulus from tensile testing with the value from modal testing, the proportional value of both testing can be approached as shown in Table 4.

Table 4. The comparison of Modulus Young value from both testing methods

Sample	E (MPa)		E tensile test/ E modal test
	Tensile Test	Modal Test	
1	43.2	30.62	1.41
2	42.9	30.64	1.40
3	44.6	30.82	1.45
4	43.2	30.30	1.43
5	42.9	30.37	1.41
AVERAGE	43.36	30.55	1.42
MIN	42.9	30.3	1.40
MAX	44.6	30.82	1.45

The average value of young modulus resulted from tensile testing is 43.36 MPa, while from modal test is 30.55 MPa. It is obtained that the average for proportional constant from the comparison of both young modulus on all data is 1,4. Therefore, the constant is added to the young modulus from modal testing to approach the value of young modulus from tensile testing. The formulation for this specific material can be written as follow:

$$E_{tensile\ testing} = E_{modal\ testing}\ x\ 1{,}4 \quad \text{... Eq. 3}$$

7. CONCLUSIONS

1. The natural frequency of materials resulted from modal testing represents the mechanical properties of materials. The differences between the two testing results in all data shows that the modal testing method is correlated with the mechanical testing and resulting the gap which is relatively consistent.
2. The average for proportional constant from the comparison of both young modulus on all data is 1,4. The formulation for this specific material is already obtained.
3. This study is conducted only on 1 (one) type of fiber-reinforced polymer in an ideal condition. It is needed to conduct further study on various kind of FRP materials, fiber orientation, aging variable, environmental condition, and other modifications.

8. REFERENCES

1. National Transportation Safety Committee, 'Final Report Ships Accident Investigation on KM. Dumai Express 10', 2009.
2. Callister, W.D, 'Material Science and Engineering, an Introduction 7th ed', 2007.
3. Crompton, T.R, 'Physical Testing of Plastics', Smithers Rapra Technology Ltd, 2012.
4. LDS-Dactron, '*Basics of Structural Testing Analysis*', 2003.
5. Agilent Technologies, 'Fundamental of signal analysis: Application Note 243, Hewlett Packard, 1994.
6. Klingenberg, Larry, 'Frequency Domain Using Excel', San Francisco State University School of Engineering, 2005.
7. Lasn, Kaspar, 'Evaluation of Stiffness and Damage of Laminar Composites', Norwegian University of Science and Technology, 2015.
8. ASTM E1876-01: Standard Test Method for Dynamic Young's Modulus, Shear Modulus, and Poisson's Ratio by Impulse Excitation of Vibration.
9. Ganci, Salvatore, 'A simple experiment on flexural vibrations and Young's modulus measurement', Research Gate, 2009.
10. Lung Cho, Chih, 'Comparison of Three Methods for Determining Young's Modulus of Wood', Taiwan J For Sci 22(3): 297-306, 2007.
11. Etcheverry, Javier et al, 'Analysis of the ASTM standards for impulse excitation of vibration and acoustic resonance techniques for rectangular parallelepipeds', VI Congreso Iberoamericano de Acústica – FIA, 2008

9. AUTHOR BIOGRAPHY

Vina Nanda holds the current position of technical staff at BKI (Biro Klasifikasi Indonesia). She has been responsible for the further study and works on the usage of ships materials, and further development of material testing in ships. Her educational background is metallurgical and materials engineering. She got both of her bachelor and master degree at University of Indonesia.

COMPARATIVE STUDY OF STRAIGHTENING METHOD USING OXY-ACETYLENE CARBURIZING WITH TUNGSTEN INERT GAS ON ALUMINUM 5083-H116

E C Septian, I K Rohmat and **M Thoriq Wahyudi,** Politeknik Perkapalan Negeri Surabaya, Indonesia

SUMMARY

This study explains how the comparison of straightening method using Oxy-Acetylene Carburizing (OAC) with Tungsten Inert Gas (TIG) at temperatures of 250-300°C and 350-400°C also cooling media using water, sea water and pneumatic wind on aluminum 5083-H116. Aluminum as cast 5083-H116 has a Yield Strength (YS) of 261.75 MPa and an elongation of 14.5%. Straightening using TIG has the optimum YS value between 218.71 and 243.59 MPa. However, the elongation value has decreased at 6.63-8.69%. Conversely, the minimum value of YS occurs on the OAC method at around 115.05-123.32 MPa. Moreover, the elongation value has increased at 15.03-19.52%. Smaller grain size on TIG has made the elongation value reduce. Al_8Mg_5 and Al_6Mn phases are able to maintain UTS, YS and hardness values. Meanwhile, the abnormal grain growth at OAC makes the elongation value raise. The massive diffusion of the oxygen element makes the UTS, YS and hardness values diminish drastically.

1. INTRODUCTION

Water transportation is one of the preferences for mobilizing, conducting trading activities, as health assistance facilities, also to maintain the boundary line of the states. The transportation is a ship. At the request of consumers who want a lightweight, fast and anti-radar ship but with an affordable price, aluminum is one of the best choice among various types of material nowadays [1]. Common material that used in aluminum shipbuilding process namely aluminum alloy 5083-H116 which belongs to the solid solution and cold work strengthening in the manufacturing process [2]. Ship construction process has applied welding process in metal joining techniques. In a case, control of the distortion during welding process becomes unavoidable. Distortion in aluminum occurs because aluminum has a higher level of thermal conductivity and linear expansion coefficient compared to steel [3]. In recent years, researchers have learned straightening techniques due to distortion of aluminum alloys welding and several of their effects [4],[5]. Straightening method using fire has been actualized in the production process owing to its low-cost and easy operation. Comparing the performance of aluminum 6N01-T5 with 7N01-T5 at different heating temperatures, [6] Suggest that suitable straightening temperatures are 200-250°C on aluminum 6N01 and 330-400°C on aluminum 7N01. This research shows that heating exceeds those temperature causes the expanding of the softening zone from quenched zone to the base metal. [7] Analysed the changes in mechanical properties and microstructure in the MIG welded joint of the 6005A aluminum alloy after straightened with different flame heating temperatures. The results showed that the hardness and tensile strength of the welded joint did not change when the heating temperature was lower than 200°C. When the heating temperature exceeded 200°C, the hardness and tensile strength decreased significantly due to grain growth and softened zone broadened. The purpose of this study is to identify temperature comparisons at 250-300°C and 350-400°C using water, sea water and pneumatic wind as cooling media in the straightening method applying oxy-acetylene carburizing and tungsten inert gas without filler metal on aluminum 5083-H116.

2. RESEARCH METHODOLOGY

2.1 MATERIALS

The material used in this study is aluminum plate 5083-H116 with dimensions of 250 x 350 x 4 mm. The chemical composition is shown in Table 1.

Table 1: Chemical composition of 5083-H116 aluminum alloy element.

Si	Fe	Cu	Mn	Mg	Cr
0.21	0.35	0.035	0.63	4.60	0.077
Ni	**Zn**	**Ti**	**Al**	***Others***	
0.0072	0.15	0.021	94.25	-	

2.2 STRAIGHTENING PROCEDURES

All straightening specimens, amounting to twelve plates, were rolled in advance to simulate distortion. Then, marking the straightening area 20 mm wide. Next, installing the wedges and clamps to hold the material to prevent free expansion due to the thermal expansion which can be observed in Figure 1. The heating was done by Z-swing in the marking area.

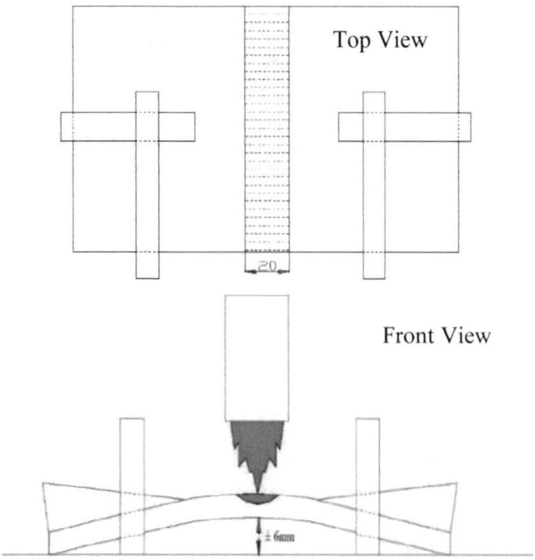

Figure 1: Marking the straightening area, wedges and clamps Installation

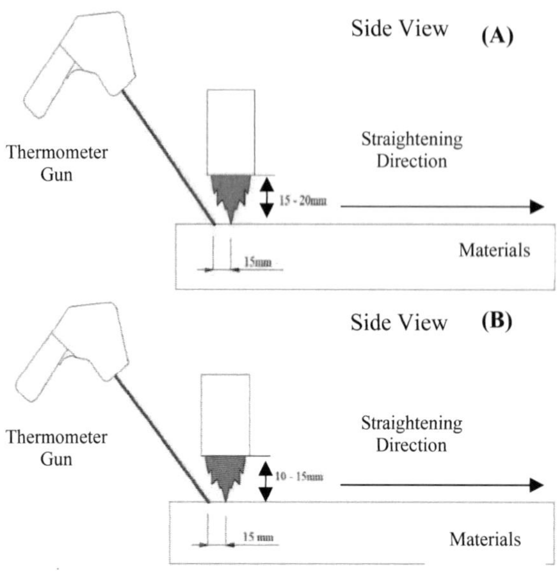

Figure 2: (A) Torch distance measurement method at 250-300°C temperature, (B) Torch distance measurement method at 350-400°C temperature

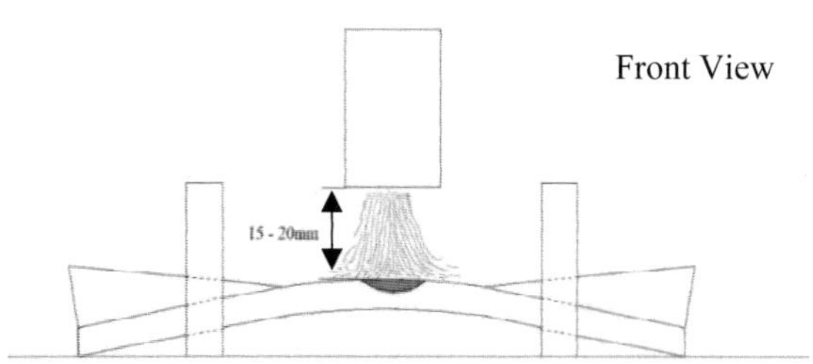

Figure 3: Cooling measurement distance

The settings for observing temperature, heating and cooling distances are illustrated in Figure 2 and Figure 3. Thermometer Gun, Stopwatch, and measured rulers are used to ensure the distance and time so that the heating and cooling processes occur constantly. There are certain parameters in the heating and cooling process which are carried out constantly: heating time for 50 seconds, cooling time of 10 seconds, cooling medium temperature of 30°C, ambient temperature of 35°C, pneumatic wind pressure of 6 Bar, working pressure of oxygen gas constant and working pressure of acetylene gas constant.

2.3 EXAMINATION PROCEDURES

There is a cutting plan for tensile, hardness and microstructure testing specimens which can be seen in Figure 4. Tensile, hardness and microstructure specimens are cut using a hydraulic cutting machine. The hardness test specimen size is 80 x 20 x 4 mm. The dimensions of the microstructure specimens are 50 x 14 x 4 mm. Tensile testing is accomplished according to the procedure in EN-ISO 6892-1: 2016. The following is the actualization of tensile specimens and determination of the location of the hardness test point in Figure 5.

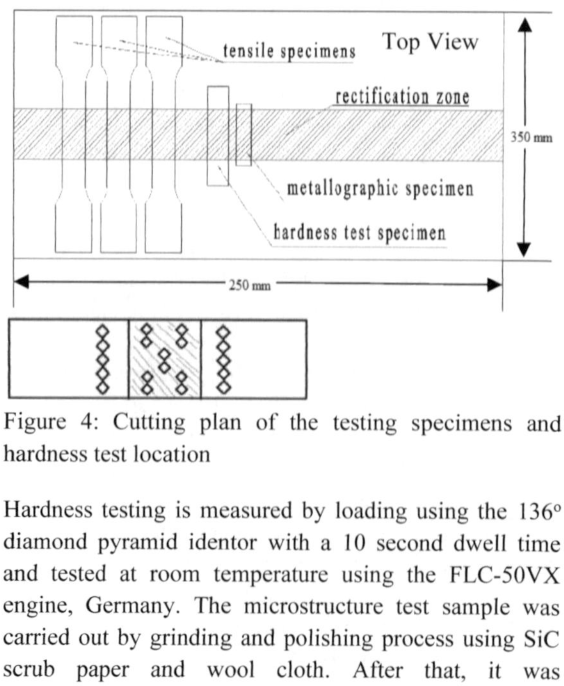

Figure 4: Cutting plan of the testing specimens and hardness test location

Hardness testing is measured by loading using the 136° diamond pyramid identor with a 10 second dwell time and tested at room temperature using the FLC-50VX engine, Germany. The microstructure test sample was carried out by grinding and polishing process using SiC scrub paper and wool cloth. After that, it was electrolyzed with H_3PO_4 for 60 seconds, then, dipped in Weck's Reagent for 30 seconds. Specimens were observed using an optical microscope from the Olympus U-25LBD, Japan. XRD testing uses Xpert Pro PANanalytical, Netherland.

3. RESULTS AND DISCUSSION

3.1 RESULTS ON TENSILE TEST, HARDNESS TEST AND COOLING RATE

Table 2: (A) Tensile test results, (B) Hardness test, cooling rate and cooling time results

Specimen	Ultimate Tensile Strength (MPa)	Yield Strength (MPa)	Elongation (%)	Failure Location
As Cast	334.5	261.75	14,5	base metal
OAC A3	275.25	119.18	16,31	base metal
OAC A4	270.6	117.23	18,33	base metal
OAC AL3	277.29	123.32	15,03	base metal
OAC AL4	271.77	118.25	17,06	base metal
OAC AP3	268.54	116.28	18,72	base metal
OAC AP4	263.48	115.05	19,52	base metal
TIG A3	284.42	239.42	7.14	straightening zone
TIG A4	277.51	231.59	7.49	straightening zone
TIG AL3	294.03	243.59	6.64	straightening zone
TIG AL4	281.5	235.75	7.36	straightening zone
TIG AP3	274.48	225.53	8.19	straightening zone
TIG AP4	266.56	218.71	8.69	straightening zone

(A)

Specimen	Hardness (HVN)	Cooling Rate (°C/minute)	Cooling Time (minute)
As Cast	116,28	-	-
OAC A3	87,09	33,33	7.95
OAC A4	84,54	35,33	10.33
OAC AL3	88,6	34,28	7.73
OAC AL4	85,24	35,85	10.18
OAC AP3	82,2	22,55	11.75
OAC AP4	76,92	26,64	13.70
TIG A3	93.23	45.14	5.87
TIG A4	88.98	54.64	6.68
TIG AL3	94.77	49.72	5.33
TIG AL4	91.80	56.85	6.42
TIG AP3	86.53	28.87	9.18
TIG AP4	79.93	30.04	12.15

(B)

Description:

- As Cast : No Treatment materials
- OAC A3 : Oxy-Acetylene Carburizing at temperature of 250-300°C using water as cooling media
- OAC A4 : Oxy-Acetylene Carburizing at temperature of 350-400°C using water as cooling media
- OAC AL3 : Oxy-Acetylene Carburizing at temperature of 250-300°C using sea water as cooling media
- OAC AL4 : Oxy-Acetylene Carburizing at temperature of 350-400°C sea water as cooling media
- OAC AP3 : Oxy-Acetylene Carburizing at temperature of 250-300°C using wind pneumatic as cooling media
- OAC AP4 : Oxy-Acetylene Carburizing at temperature of 350-400°C using wind pneumatic as cooling media
- TIG A3 : Tungsten Inert Gas at temperature of 250-300°C using water as cooling media
- TIG A4 : Tungsten Inert Gas at temperature of 350-400°C using water as cooling media
- TIG AL3 : Tungsten Inert Gas at temperature of 250-300°C using sea water as cooling media
- TIG AL4 : Tungsten Inert Gas at temperature of 350-400°C sea water as cooling media
- TIG AP3 : Tungsten Inert Gas at temperature of 250-300°C using wind pneumatic as cooling media
- TIG AP4 : Tungsten Inert Gas at temperature of 350-400°C using wind pneumatic as cooling media

In straightening using OAC, the tensile and hardness test results show that both of these tests have a value that is directly proportional. However, inversely proportional to the elongation value, cooling rate and cooling time. The value of ultimate tensile strength and yield strength decreased significantly when compared to As Cast material. Meanwhile, the elongation value increases alongside increasing of temperature. cooling rate and longer cooling time. Straightening using TIG shows a phenomenon that is quite different from OAC. The value of ultimate tensile strength and yield strength reduces. However, the reduction that occurred was not too drastic. Conversely, the value of elongation has decreased by half the value of As Cast. Overall, the hardness value in both straightening methods declined significantly when the heating temperature rose from 250-300°C to 350-400°C. This phenomenon is due to the characteristic of 5083 aluminum material which has a low recrystallization temperature of 240°C, this is the temperature in which nucleation and grain growth occur. Crystals will undergo nucleation in areas of dislocation with high density and therefore tend to experience softening in the material. Nuclei grow beside time and form on the aluminum matrix. Thus, the grains will touch each other. In addition, full anneal temperatures are low at 345°C so as to cause permanently softening and a loss in the strength value of aluminum 5083 [8],[9],[10]. The cooling media factor has a better influence on the cooling mechanism in the water and sea water media so that it can provide a faster quenching effect and maintain the strength value of aluminum material 5083 [11]. Several factors that influence cooling become faster are: heating temperature, viscosity and density of quenchants [12],[13]. [14] States that if the density is greater and the viscosity is smaller, then, the faster the cooling rate will occur.

3.2 MICROSTRUCTURE RESULTS

In general, As Cast material has an elongated grain and there are precipitates at each grain boundary and partially spread in grain. This is owing to the aluminum material 5083-H116 is an aluminum alloy that obtains the grain shape and strengthening mechanism from the solid solution strengthening process by adding main alloying elements such as magnesium, manganese, iron and silicon. Furthermore, a cold working process is carried out which causes the grains to form flat and elongated. The combination of solid solution strengthening and cold working is then said to be precipitation and strain hardening which are able to inhibit dislocation in Aluminum 5083-H116 when deformation occurs [10],[15]. The precipitates in Figures 5 (A) and (B) are shown by black arrows. Meanwhile, elongated grain is indicated by red arrows. The microstructure of OAC AL3 looks different from the As Cast material. The Elongated Grain shape has disappeared and has been replaced by grain shapes that look like small blocks. Subgrain is easily formed on aluminum material 5083-H116 because it has a low recovery temperature at 195°C.

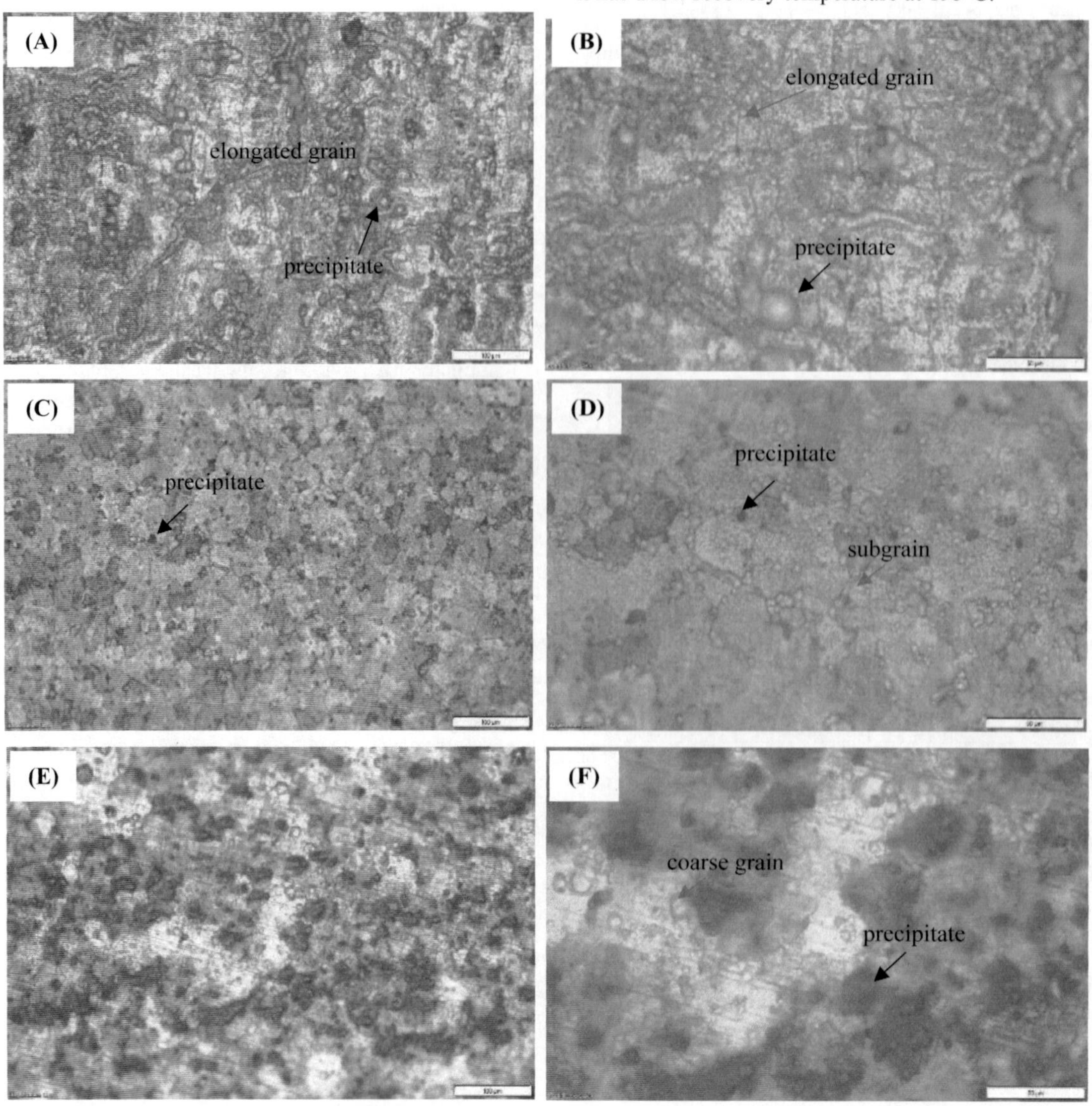

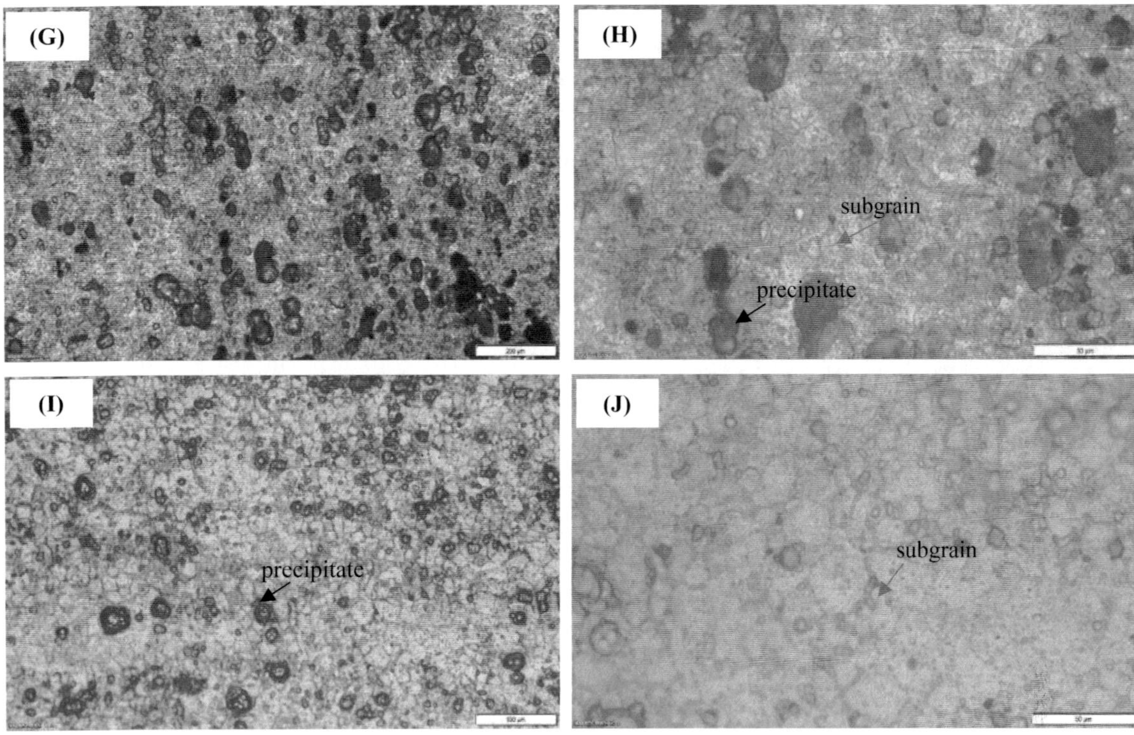

Figure 5: (A) As Cast Microstructure, (C) OAC AL3, (E) OAC AP4 (G) TIG AL3, (I) TIG AP4 – 200x magnifying, (B) As Cast, (D) OAC AL3, (F) OAC AP4 (H) TIG AL3, (J) TIG AP4 – 500x magnifying

The formation of subgrain which grows and binds to each other will become subboundaries that make the dislocation experience movement and rearrangement due to thermal activation into a low energy configuration. Of course, it will experiences more annihilation [10]. Subgrain is seen more clearly as shown by the red arrows in Figure 7 (C) and (D). Precipitates begin to disperse evenly and are smaller at each grain boundary. The precipitates in the microstructure in Figures 7 (C) and (D) are marked by black arrows. The grain size of the OAC AP4 appears to increase significantly, the microstructure is very contrasting from the As Cast and OAC AL3 variations. The elongated grain shape also disappears and tends to form round and coarse grains as shown by the red arrows in Figure 7 (E) and (F). These round and rough grains can be said to be an abnormal grain growth phenomenon. Simultaneously, the black precipitate begins to enlarge and multiply is shown in black arrows in Figure 7 (E) and (F). Precipitates tend to gather and try to tie each other. This happens because of the protractedly cooling rate so that the dissolved elements lead supersaturated conditions in the aluminum matrix. These elements namely magnesium and manganese which diffuse into new precipitates and bind to other elements that can cause a decrease in the effect of strengthening the solid solution on the As Cast material. In the TIG AL3 straightening method from Figure 5 (G) and (H), the microstructure is different in

In areas affected by direct heat exposure, subgrain formation occurs as indicated by the red arrows. However, the straightening area is still able to maintain a large precipitate size and tend to cluster. This shows that precipitates that appear to have a size that is not much different are relatively able to maintain the value of UTS, YS and the value of hardness to be better. However, the grain size is smaller and causes the elongation value to decrease compared to As Cast material. It can clearly be seen from Figure 5 (I) and (J), that the microstructure in the TIG AP4 variation has a larger grain size when compared to the TIG AL3 specimen. The formation of much subgrains, as shown by the red-colored arrows occurs due to the process of heat exposure with higher temperatures and slow cooling rates [16]. The shrinking precipitate is shown in black arrows and spread evenly at the grain boundaries and the formation of subgrain which results in decreased UTS, YS values and hardness. On the contrary, this straightening process is able to maintain a few elongation value.

3.3 X-RAY DIFFRACTION (XRD) RESULTS

As Cast XRD test result provide similar outcome according to [17],[18],[19]. Phases contained in aluminum material 5083, namely: α Aluminum, β (Al_8Mg_5), Mg_2Si and $Al_{12}(FeMn)_3Si$.

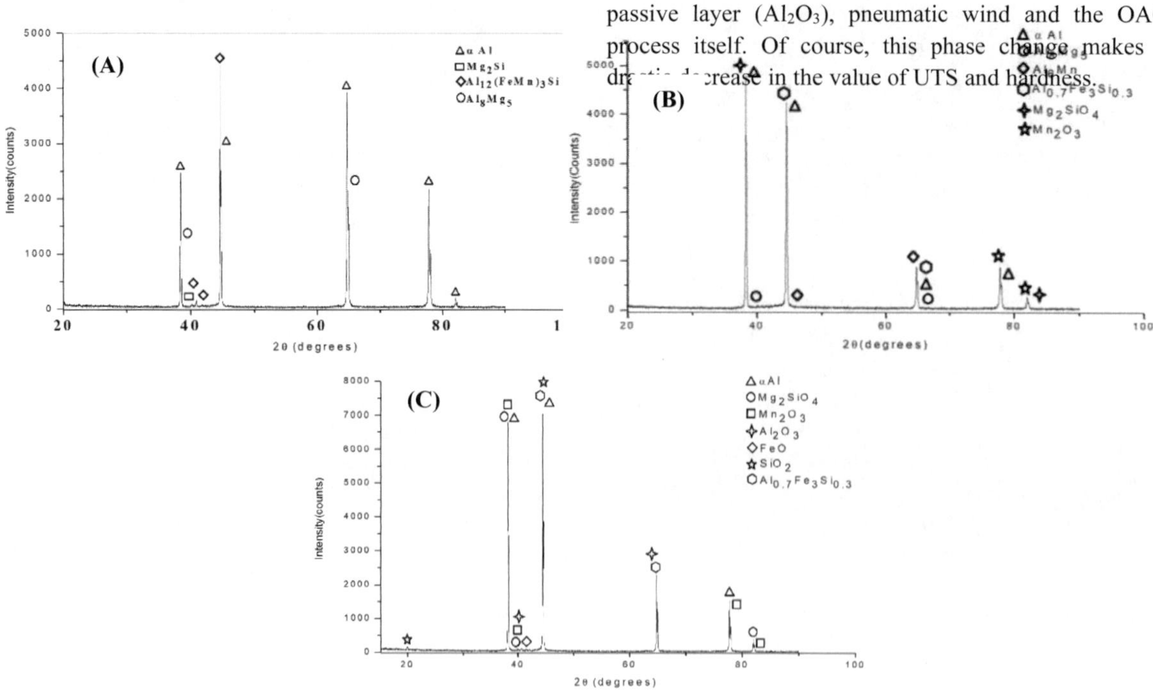

Figure 6: (A) XRD As Cast, (B) XRD TIG AL3, (C) XRD OAC AP4

It can be seen that in the XRD test on TIG AL3 shown the tendency of a compound to bind to other compounds that have an impact on the main phase changes in the As Cast material. The Mg_2Si phase binds Oxygen by the interstitial impurities mechanism into Mg_2SiO4, $Al_{12}(FeMn)_3Si$ separates into three distinct phases, namely Al_6Mn, $Al_{0.7}Fe_3Si_{0.3}$ with substitution mechanism. Unlike Mn_2O_3 which has the same mechanism as Mg_2SiO_4. The oxygen that diffuses in the TIG AL3 specimen is obtained from the passive layer (Al_2O_3) which is not grinded when the straightening process is carried out and sea water also has oxygen content. Three phases in 5083 aluminum alloy, namely α Aluminum, Al_8Mg_5 and Al_6Mn are still in the TIG AL3 matrix material. This shows that the rapid cooling rate and temperature that has not yet entered the grain growth phase is able to maintain these two main phases in order to stabilize the value of UTS, YS and hardness [20]. The mechanism of oxygen interstitial impurities in OAC AP4 in Figure 8 occurs very massive. This is caused by the heating temperature and the long time cooling rate which causes the material to undergo grain growth which results in annihilation of the matrix so that diffusion occurs through two mechanisms [21]. The microstructure and XRD tests on the OAC AP4 specimens showed that this oxygen diffusion was able to make the aluminum alloy 5083 undergo a phase change. Almost all phases in As Cast material change and bind to oxygen. Oxygen which is a nonmetallic inclusion is obtained from the passive layer (Al_2O_3), pneumatic wind and the OAC process itself. Of course, this phase change makes a drastic decrease in the value of UTS and hardness.

4. CONCLUSIONS

Temperature and cooling media give different results to the cooling rate in the straightening variation in this study. Temperature 250-300°C and seawater cooling media are variations that have the fastest cooling rate. Meanwhile, temperatures of 350-400°C and pneumatic wind cooling media are variations that have the slowest cooling rates. As Cast aluminum 5083-H116 has a UTS value of 334.5 MPa, YS of 261.75 MPa, hardness of 116.26 HVN and 14.5% elongation. The straightening method using TIG has decreased UTS, YS values and hardness not too significant. While the value of elongation decreased in the range of 6.63-8.69%. The straightening method using TIG tends to maintain the Al_6Mn and Al_8Mg_5 phases on the 5083-H116 aluminum alloy so that it is able to withstand the degradation of UTS, YS and hardness values. However, smaller grain sizes cause the elongation value to decrease. The slow cooling rate causes the UTS and hardness values to reduce in the range of 263.48-277.29 MPa and 76.92-88.6 HVN. Moreover, the value of YS fell at 115.05-123.32 MPa. However, elongation increased at 15.03-19.52%. Abnormal grain growth in OAC rises the value of elongation. Massive diffusion of the oxygen element forms new compound bonds so that the value of UTS, YS and hardness losses drastically.

5. ACKNOWLEDGEMENTS

This work was supported by PT PAL INDONESIA (PERSERO), Characterisation Laboratory (Material and Metallurgy Department, Institur Teknologi 10 Nopember Surabaya) and DT-NDT Laboratory (Politeknik Perkapalan Negeri Surabaya).

6. REFERENCES

1. Xue, J., Liu, J., 'Research on current state and prospect of welding deformation and deformation control methods for China's high-speed train aluminum alloy car body', *Hot Work. Technol. 41, 201–208*, 2012.
2. ASM International, 'Aluminum and Aluminum Alloys', *ASM International*, 2001.
3. BOC Group, 'Fundamentals of Flame Straightening', *BOC Group*, 2009.
4. Chen, X. L., 'Corrosion Behavior, Yield Behavior and High Temperature Deformation Behavior of Marine High Magnesium Aluminum Alloy', *Master's Thesis. Central South University*, 2009.
5. Li, S., Guo, D., Dong, H.D., 'Effect of flame rectification on corrosion property of Al-Zn-Mg alloy', *Trans. Nonferrous Met. Soc. 27, 250–257*, 2016.
6. Xiong, Z. L., 'Effect Mechanism of Heat-Straightening Temperature on Microstructure and Properties of Aluminum Alloy Joint in High-Speed Trains', *Master's Thesis. Harbin. Harbin Institute of Technology*, 2014.
7. Jiang, L., Wei, X. J., Yao, G. C., Wang, D. Q., 'Effect of Flame Heating Temperature on Properties of Welded Joint of 6005A Aluminium Alloy', *J. Northeast. Univ, 24, 365–368*, 2003.
8. I-Car, 'Straightening Aluminum Program 1', *Inter-Industry Conference On Auto Collision Repair*, 2002.
9. Lumley, R., 'Fundamentals of Aluminum Metallurgy', *Woodhead Publishing Limited*, 2011.
10. Totten, George, E., 'Volume 4E Heat Treating of Nonferrous Alloys', *ASM International*, 2016.
11. Cai, Congwei., Wang, Xue., Liang, Zhimin., Rao, Yuzhong., Wang, Hongbo, Yan, Dejun., 'Effects of Water Cooling on the Mechanical Properties and Microstructure of 5083 Aluminum Alloy during Flame Straightening', *MDPI*, 2018.
12. Kalpakjian, S., Schmid, Steven R., 'Manufacturing Engineering and Technology Sixth Edition in SI Units', *Pearson Prentice Hall*, 2009.
13. Lisci´c, Bozidar., Tensi, Hans M., Canale, Lauralice C. F., Totten, George E., 'Quenching Theory and Technology', *CRC Press*, 2009.
14. Rajan, T. V., Sharma, C. P., Sharma, Ashok, 'Heat Treatment: Principles and Techniques', *Prentice Hall of India Private Limited*, 2011.
15. Anderson, Kevin., Weritz, John., Kaufmann, J. Gilbert, 'Volume 2A. Aluminum Science and Technology', *ASM International*, 2018.
16. Sheppard, T., Parson, N.C., Zaidi, M. A., 'Dynamic recrystallization in Al-7Mg Alloy', *Department of Metallurgy and Materials Science – Imperial College of Science and Technology*, 1983.
17. Belov, Nikolay A., Eskin, Dimitry G., Aksenov, Andrey A., 'Multicomponent phase diagrams Application for commercial aluminum alloys', *ElSevier*, 2005.
18. Shankar, Khrisnakumar., Wu, Weidong., 'Effect of welding and weld repair on crack propagation behaviour in aluminium alloy 5083 plates', *ElSevier*, 2001.
19. Kuijpers, Niels C. W., Vermolen, Fred J. Vuik., Kees, Sybrand van der Zwaag., 'A Model of the AlFeSi to Al(FeMn)Si Transformation in Al–Mg–Si Alloys', *The Japan Institute of Metals*, 2003.
20. Martienssen, Professor Werner., Warlimont, Professor Hans., 'Springer Handbook of Condensed Matter and Materials Data', *Springer Berlin Heidelberg*, 2005.
21. Chen, Ren Yu., Chu, Hsiao Yeh., Lai, Cheng Chyuan., Wu, Chih Ting., 'Effect of annealing temperature on the mechanical properties and sensitization of 5083-H116 aluminum alloy', SAGEPUBLICATIONS, 2013.

7. AUTHORS BIOGRAPHY

Eka Cahya Septian, student at Politeknik Perkapalan Negeri Surabaya. He was responsible for analysed and did the testing on specimens.

Imam Khoirul Rohmat, lecturer at Politeknik Perkapalan Negeri Surabaya. He was responsible for analysed and writing content.

Mohammad Thoriq Wahyudi, lecturer and Head of DT-NDT Laboratory at Politeknik Perkapalan Negeri Surabaya. He was responsible for advised and did the testing on specimens.